STOPOVER ECOLOGY OF NEARCTIC-NEOTROPICAL LANDBIRD MIGRANTS: HABITAT RELATIONS AND CONSERVATION IMPLICATIONS

Frank R. Moore, editor

Sponsors:

Gulf Coast Bird Observatory
Houston Audubon Society
U.S.D.A. Forest Service, Rocky Mountain Research Station
University of Southern Mississippi

Studies in Avian Biology No. 20
A PUBLICATION OF THE COOPER ORNITHOLOGICAL SOCIETY

Cover drawing of Gray Catbirds (*Dumetella carolinensis*) winging over the Gulf of Mexico by Michelle Davis.

STUDIES IN AVIAN BIOLOGY

Edited by

John T. Rotenberry
Department of Biology
University of California
Riverside, CA 92521

Studies in Avian Biology is a series of works too long for *The Condor,* published at irregular intervals by the Cooper Ornithological Society. Manuscripts for consideration should be submitted to the editor. Style and format should follow those of previous issues.

Price $18.00 including postage and handling. All orders cash in advance; make checks payable to Cooper Ornithological Society. Send orders to Cooper Ornithological Society, % Western Foundation of Vertebrate Zoology, 439 Calle San Pablo, Camarillo, CA 93010.

ISBN: 1-891276-12-3

Library of Congress Catalog Card Number: 99-080020
Printed at Allen Press, Inc., Lawrence, Kansas 66044
Issued: 7 January 2000

CONTENTS

LIST OF AUTHORS

DAVID A. ABORN
Department of Biological Sciences
University of Southern Mississippi
Hattiesburg, MS 39406
(Present address: Department of Biological and Environmental Sciences
University of Tennessee at Chattanooga
Chattanooga, TN 37403)

WYLIE C. BARROW, JR.
U.S. Geological Survey
National Wetlands Research Center
700 Cajundome Blvd.
Lafayette, LA 70506

CHAO-CHIEH CHEN
School of Forestry, Wildlife, and Fisheries
Louisiana State University
Baton Rouge, LA 70803
(Present address: Institute of Statistical Science
Academia Sinica
Taipei 11541, Taiwan)

DEBORAH M. FINCH
USDA Forest Service
Rocky Mountain Research Station
2205 Columbia SE
Albuquerque, NM 87106

ROBERT B. HAMILTON
School of Forestry, Wildlife, and Fisheries
Louisiana State University
Baton Rouge, LA 70803

RICHARD L. HUTTO
Division of Biological Sciences
University of Montana
Missoula, MT 59812

SARAH E. MABEY
Department of Conservation and Recreation
Division of Natural Heritage
1500 E. Main Street
Richmond, VA 23219
(Present address: Department of Biological Sciences
University of Southern Mississippi
Hattiesburg, MS 39406)

FRANK R. MOORE
Department of Biological Sciences
University of Southern Mississippi
Hattiesburg, MS 39406

KEITH OUCHLEY
The Nature Conservancy
P. O. Box 4125
Baton Rouge, LA 70821

JEFFREY DAVID PARRISH
Department of Ecology and Evolutionary Biology
Box G-W
Brown University
Providence, RI 02912
(Present address: Caribbean Division
The Nature Conservancy
4245 No. Fairfax Drive
Arlington, VA 22203)

SCOTT M. PEARSON
Department of Biology
Mars Hill College
Mars Hill, NC 28754

DANIEL R. PETIT
U.S. Geological Survey
Biological Resources Division
12201 Sunrise Valley Drive
Reston, VA 20192

THEODORE R. SIMONS
Cooperative Fish and Wildlife Research Unit
Department of Zoology
North Carolina State University
Raleigh, NC 27695

TERRY J. SPENGLER
U.S. Geological Survey
National Wetlands Research Center
700 Cajundome Blvd.
Lafayette, LA 70506

BRYAN D. WATTS
Center for Conservation Biology
College of William and Mary
Williamsburg, VA 23187

MARK S. WOODREY
Department of Biological Sciences
University of Southern Mississippi
Hattiesburg, MS 39406-5018
(Present address: Mississippi Museum of Natural Science
Mississippi Department of Wildlife, Fisheries and Parks
111 North Jefferson St.
Jackson, MS 39201)

WANG YONG
USDA Forest Service
Rocky Mountain Research Station
2205 Columbia SE
Albuquerque, NM 87106
(Present address: Department of Natural Resource Science
University of Rhode Island
Kingston, RI 02881)

Studies in Avian Biology No. 20:1–3, 2000.

PREFACE

FRANK R. MOORE

Each year billions of landbirds migrate between the northern and southern hemispheres of both the New and Old World. In eastern North America alone, over two thirds of all the breeding bird species migrate from temperate breeding grounds to more tropical wintering areas in the Caribbean, Mexico, and Central and South America. The benefits of intercontinental migration, regardless of whether they accrue through increased survivorship by overwintering in the tropics, increased productivity by breeding in seasonally rich temperate areas, or both, must be balanced against costs of migration. Traveling long distances between temperate and tropical areas comes with considerable risks, and the mortality associated with intercontinental migration, though difficult to estimate, may be substantial. Consider some of the problems a migrant faces during passage, not the least of which is the energetic cost of transport. Migrants must also adjust to unfamiliar habitats, conflicting demands between predator avoidance and food acquisition, competition with other migrants and residents for limited resources, unfavorable weather, and orientation errors. To the extent migrants solve those problems they experience a successful migration, one measured ultimately in terms of survival and reproductive success.

The long-distance movements and biology of migratory birds during stopover has generated considerable interest in recent years, in no small part because of threats to their populations. Although reports of drastic declines for the group as a whole are exaggerated, some migrant landbirds are showing long-term population declines. Decline in populations has been attributed to events on the wintering grounds, fragmentation of breeding habitat, and to changes in the suitability of *en route* (stopover) habitat. For a Red-eyed Vireo or a Yellow-billed Cuckoo, the choice of habitat must be made in tropical wintering quarters, temperate breeding areas, and repeatedly during migration. Consequently, factors associated with the stopover ecology of migrants must figure in any analysis of population change and in the development of a comprehensive conservation "strategy" for landbird migrants. Protect all the breeding woodland in North America and all of the appropriate habitat on the wintering grounds and populations of intercontinental migrants will still decline unless habitat requirements during migration are factored into the conservation equation.

The contributions to this issue of *Studies in Avian Biology* focus on migrant-habitat relations during passage and on the conservation implications of that relationship. Few migratory birds engage in nonstop flights between points of origin and destination; rather they stopover periodically—they land for a few hours or a few days before resuming migratory flight. A stopover site is any place where a migratory bird pauses for some length of time between migratory flights. What is the value of a stopover site for a migrating bird? What factors determine the quality of a particular stopover site? The answer to those non-trivial questions depends on understanding the migrant's relationship to habitat.

When contemplating the stopover ecology of migratory birds, it is essential to recognize that migration occurs over a broad geographic scale, but over a relatively short temporal scale, and that a migrating bird's relationship to habitat is scale-dependent (i.e., different factors, some extrinsic to habitat *per se,* operate at these different scales). Intrinsic constraints on habitat use are those factors thought to determine habitat quality and upon which migrants made decisions about habitat use (e.g., food, presence of predators). As the spatial scale broadens, factors intrinsic to habitat give way to factors largely unrelated to habitat (extrinsic constraints), such as synoptic weather patterns during passage. The study of the landbirds during migration should reflect the hierarchical nature of the migrant's relationship to habitat. In the first contribution to this issue, Ted Simons and his colleagues ask us to step back and view this relationship at the landscape scale. The movement of birds across the Gulf of Mexico each spring and fall provides the geographical context for application of spatially explicit models to the stopover of landbird migrants.

Daniel Petit asks what types of habitat are important to migrating songbirds when they pause during passage. Over the course of a season's migration, a migratory bird encounters a variety of habitats, most of them new habitats with associated new food, new competitors, and new predators. After a night's passage it finds itself in a habitat that may be very different from the one occupied the previous day, let alone the previous year. Moreover, favorable *en route* habitat, where migrants can rapidly accumulate energy stores, is probably limited in an absolute sense, or effectively so because migrants have limited time to search for the "best" stopover site. Nevertheless, evidence indicates that migrants prefer certain habitats and select among alternatives during stopover, presumably in response to differential suitability. Suitability of *en route* habitat depends largely on three factors: (1) foraging opportunities, (2) competition with other migrants and with residents, and (3) shelter against predators and adverse weather. Beyond those generalities, our understanding of the determinants of habitat suitability is not very refined and open to speculation.

Whereas evidence reveals that habitat selection occurs during migration, little is known about how migrants made decisions about habitat use during stopover. David Aborn and I ask about the mechanisms of habitat selection: How do migrants distinguish one habitat from another? How is habitat quality assessed? What cues do migrants use when deciding to settle in a particular habitat? We are only beginning to understand migrant-habitat relations during migration, much less appreciate the mechanisms migrants use to identify habitat attributes on which habitat choices are made during passage.

Mark Woodrey calls attention to age-dependent aspects of stopover biology. If the high cost of migration (i.e., reduced fitness; increased mortality) is absorbed largely by inexperienced, hatching-year birds, differential costs should be reflected in age-dependent differences in stopover biology. Presumably yearling migrants experience more trouble solving *en route* problems than older, more experienced migrants. What is the empirical basis for this supposition? Exactly which problems are most likely to create an age-dependent consequence? Moreover, individuals with different levels of migratory experience can be expected to respond differently to the exigencies of migration.

Migration is an energetically demanding task, and fat is the essential source of energy to fuel migratory flights. In anticipation of the energetic demands of migration, birds become hyperphagic and deposit as much as 50% of the normal body mass in fat stores. For intercontinental migrants the energy requirements necessary to reach their destination exceed even this amount several times over, so migrant landbirds stop periodically to rest and refuel. Although it seems obvious that the single most important constraint during migration is to acquire enough food to meet energetic requirements, satisfying energy demand is not simply a matter of hyperphagia. The availability of nutrients specific to a particular need, such as calcium in relation to egg formation for females during spring migration or certain fruits that facilitate fat deposition, must be taken into account when considering food availability. Such constraints could affect not only the rate at which migrants replenish energy stores, but also the migrant's susceptibility to predator attack. Jeffrey Parrish examines the dietary flexibility of migratory birds during passage and the conservation implications of food choice.

The coastal woodlands and narrow barrier islands that lie scattered along the northern coast of the Gulf of Mexico provide important stopover habitat for landbird migrants. They represent the last possible stopover before fall migrants make an 18–24 hr, nonstop flight of greater than 1,000 km, and the first possible landfall for birds returning north in spring. Yet, the northern coast of the Gulf of Mexico is experiencing significant human population increases and concomitant development. The southward migration of industry coupled with changing demographics will increase pressure on stopover habitats in the decades ahead. As stopover habitat is transformed or degraded and the cost of migration increases, there is a commensurate increase in the value of unaltered habitat to migratory birds, which makes the creation of new habitats to replace those lost to coastal development a major conservation challenge in the next century. Wylie Barrow and his colleagues address restoration of stopover habitat in relation to the chenier plain of southwestern Louisiana.

Information on the spatial and temporal pattern of migration, not to mention migration volume ("traffic rate"), is not readily available for the southwestern United States or the West in general. Yet, it is clear that riparian or riverine habitats in the southwestern United States are vital to landbird migrants, notably woodland species. Deborah Finch and Wang Yong examine the vegetational and human history of the middle Rio Grande River in relation to its importance to landbird migrants during passage. Their contribution prompts us to recognize that corridors of riparian habitat may represent critical stopover areas regardless of geographical region.

The spatial scale over which migration occurs coupled with the variety of habitats migrants encounter during passage made the challenge of conserving stopover habitat for landbird migrants uniquely different from that of protecting breeding or wintering habitats. Sarah Mabey and Brian Watts correctly point out that most conservation strategies focus on large tracts of public and private lands. What of threats on the aggregate of relatively small, private land parcels? The authors describe the use of policy and management tools that take us beyond the boundaries of public land and illustrate their application on the lower Delmarva Peninsula, Northhampton County, Virginia.

In the closing contribution, Richard Hutto calls attention to several issues, some peculiar to the migratory period, that are important to the conservation of landbird migrants: (a) patterns of geographic distribution during passage, (b) patterns of habitat use during passage, (c) stopover events in relation to population regulation, and (d) the story-telling power of migration. He reminds us that the success of our conservation efforts is tied to our attitudes about conservation. Our fascination with the sheer drama and beauty of the migratory journey contributes tangibly to the development of a conservation ethic.

I am especially grateful to John Rotenberry for his patience, persistence, and editorial efforts. Many colleagues, including Robert Caldow, David Cimprich, Robert Cooper, Brent Danielson, Dave Ewert, John Faaborg, Rebecca Holberton, Chuck Hunter, Richard Hutto, Paul Kerlinger, Tom Litwin, Kathy Milne, David Pashley, Tom Sherry, and Charles Smith, contributed to the publication of this issue through their careful, constructive reviews of different contributions. Support toward publication of this issue of *Studies in Avian Biology* was generously provided by the Gulf Coast Bird Observatory, the Houston Audubon Society, the USDA Forest Service Rocky Mountain Research Station, and the University of Southern Mississippi.

Studies in Avian Biology No. 20:4–14, 2000.

APPLICATION OF SPATIAL MODELS TO THE STOPOVER ECOLOGY OF TRANS-GULF MIGRANTS

THEODORE R. SIMONS, SCOTT M. PEARSON, AND FRANK R. MOORE

Abstract. Studies at migratory stopover sites along the northern coast of the Gulf of Mexico are providing an understanding of how weather, habitat, and energetic factors combine to shape the stopover ecology of trans-Gulf migrants. We are coupling this understanding with analyses of landscape-level patterns of habitat availability by using spatially explicit models to simulate avian movements through stopover habitats. The probability that an individual migrant will complete a migration successfully is determined by the bird's energetic status and flight morphology, and the quality, quantity, and spatial pattern of habitats encountered during migration. The models evaluate habitat patches according to their distance from the coast, isolation from other patches of suitable habitat, and habitat quality. Evaluation procedures have been developed from available data on the arrival condition of migrants, energetic and morphological constraints on movement, and species-specific habitat preferences. Window analysis and individual-based modeling are used to demonstrate how the abundance, quality, and spatial pattern of habitats interact with the arrival energetic state of migrants to determine the suitability of migratory stopover habitats along the northern Gulf coast. Our goal is to understand how landscape-scale patterns of habitat conversion may be affecting populations of trans-Gulf migrants.

Key Words: birds, landscape pattern, migration, spatial models, stopover ecology.

Ecologists are beginning to appreciate how the spatial and temporal scale of the data they collect influence their understanding of natural patterns and processes (Wiens 1981, 1989; Edwards et al. 1994, Pearson et al. 1996). As May (1994) has recently pointed out "the answers to ecological questions—and ultimately the understanding of ecological systems—depend on whether or not the system is studied at an appropriate scale," noting an "increasing need for ecologists in general, and conservation biologists in particular, to deal with larger spatial scales than most of us are used to, or happy with."

Recent declines in populations of nearctic-neotropical landbird migrants (Robbins et al. 1989b, Askins 1990) have prompted a wave of new research into the factors affecting populations of these birds on their breeding and wintering grounds (Hagan and Johnson 1992, Finch and Stangel 1993) and a smaller number of studies on the factors affecting birds during migration (Moore and Simons 1992a, Watts and Mabey 1993, Moore et al. 1995). Designing conservation-oriented studies of the stopover ecology of migrants is complicated by the fact that migration occurs over a broad geographic scale, but over a relatively short temporal scale.

Remote sensing technology and spatial modeling techniques are providing new research tools for investigating how the distribution and abundance of habitats may be affecting wildlife populations. Our objective is to use these tools to understand how variation in the landscape-level pattern of habitats affects migrant birds. We will use spatially explicit models to explore the effects of changing landscape patterns on the probability of a successful migration. These models, while simplistic, incorporate some basic bird biology and analyze landscape-level variation in habitats from the perspective of migrants with different energetic states. We hope that the results of this analysis will be useful in setting priorities for future research and conservation.

The conceptual framework for developing our spatial models is straightforward (Fig. 1). Spring migrants make landfall in landscapes containing habitats that vary in suitability for foraging. The abundance and spatial pattern of high-quality habitat in these landscapes will likely affect the probability of a successful migration. We know that arriving migrants vary in their energetic condition—some are lean, while some have considerable fat stores remaining. As long as favorable habitat is readily available, both fat and lean birds eventually find suitable habitat. But as suitable habitat is lost and accessibility declines, a fat-depleted migrant's ability to find good habitat may be limited because the benefits of rejecting suboptimal habitat may be outweighed by the cost of finding better sites. Ultimately, the interplay of a migrant's energetic state and the abundance and spatial configuration of stopover habitats, will determine the likelihood of a successful migration.

METHODS

Landscape-level metrics provide a means to quantify the abundance and spatial pattern of habitat types in study landscapes (Turner and Gardner 1991). The most straight-forward measure is the area of suitable habitat types. Habitat connectivity or fragmentation can also be measured using indices of spatial pattern. Examples of such indices include contagion (the probability that

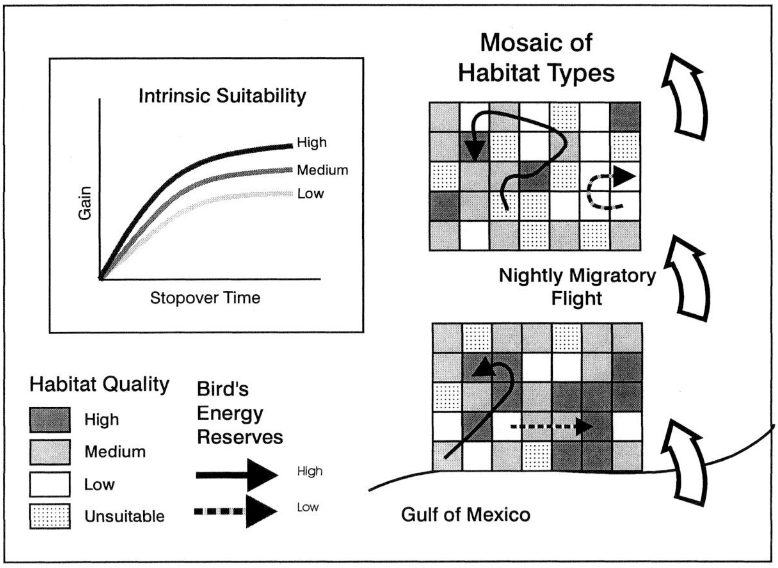

FIGURE 1. Conceptual spatial model. Migrants arrive along the northern Gulf coast with different amounts of stored fat, and they encounter habitats of varying intrinsic suitability. When high quality stopover habitat is available (lower matrix) birds with both high and low energy reserves find suitable stopover habitat. As suitable habitat is lost (upper matrix) birds begin to use sub-optimal stopover sites, which may reduce the probability of a successful migration, especially for birds with low energy reserves.

two adjacent cells are of the same habitat type), the number and size of patches of each habitat type, and the area of the largest patch divided by the total area of all patches of that habitat type. This final index provides a measure of fragmentation that varies over the interval [0,1] where 0 = highly fragmented and 1 = a homogeneous landscape. These metrics provide a means to quantitatively compare landscapes. The models described below provide measures of landscape conditions from the perspective of migrant birds. These models include (1) a window analysis that assesses the landscape in the vicinity of a bird making landfall, and (2) an individual-based model that simulates the energetic state of birds foraging in habitats of varying quality.

Model Input Parameters

The parameters in our models included energetic, flight performance, and habitat variables. The energetic status of spring migrants was measured between 1987–1994 using mist nets to sample birds at stopover sites along the northern Gulf coast (Moore et al. 1990, Kuenzi et al. 1991, Moore and Simons 1992a). Birds were weighed on electronic scales to the nearest 0.05 gram, banded, and released. Fat reserves were estimated by visual inspection of all birds, which were ranked on an ordinal scale from zero to five according to the method described by Helms and Drury (1960).

Measurements of birds' energy reserves and wing spans were used to calculate flight range estimates, using the flight performance equations developed by Pennycuick (1989).

Habitat data were derived from a supervised classification of two 1990 Landsat Thematic Mapper scenes of the northern Gulf coast produced by the National Biological Service Southern Science Center in Lafayette, LA. This map was comprised of 18 original cover types in raster format, with a cell size of 28.5 m × 28.5 m. The 18 original cover types were aggregated to produce four habitat types that were then used in all spatial analyses (see RESULTS).

The habitat associations of birds were determined through a combination of 10-min point counts (N = 500 points) at barrier island sites (Moore et al. 1990) and 1-km strip transects (Emlen 1977) at mainland sites (N = 117 transects from 9 paired sites, see Table 2 for sampling design; Moore and Simons 1992b). Census results were then used to assign each of the original 18 habitat types to one of four habitat categories that ranged from low (category 1) to high (category 4) suitability as migratory bird stopover habitat. These four habitat categories were used in all subsequent analyses. This ranking of habitat quality assumes that the relative abundance of migrants in stopover habitats reflects relative habitat quality although this assumption was not tested empirically.

Spatial Analyses

We used spatial analyses to examine how the abundance and spatial configuration of habitats might affect the suitability of stopover habitat for spring migrants. We did this using a window analysis technique and through the application of an individual-based model to our field data and habitat map.

Window analysis

In the window analysis, a hypothetical individual bird was randomly located in a block of arrival habitat.

A window was then projected from the arrival location, with the size of the window reflecting the individual bird's energetic state. This window represented the area that could be searched and sampled by a bird, given its energetic condition on arrival (i.e., the greater the bird's energy stores, the larger the window). Habitat measures, such as mean habitat rank, were calculated from all of the cells within a window. The window's pie-piece-like shape reflected a migrant's tendency to move northward during spring migration (Gauthreaux 1991). The window analysis allowed us to quantify the range of foraging conditions experienced by arriving birds, and the probability that a single bird would land in an area of specified quality (e.g., very rich, moderate, or poor quality).

Individual-based model

A second approach involved the development of an individual-based model. This method allowed us to begin to examine the relative importance of and the interaction between the energetic state of arriving birds and the spatial pattern of habitat within a landscape. It is impossible to precisely model the details of the behavior and energy dynamics of birds during stopover because of our lack of data and knowledge about these organisms. However, this model incorporates the most basic components of the biology of a migrant: (a) variation in habitat quality, and (b) changes in its energetic state due to foraging.

Our model used an Energy State Index (ESI) to indicate the relative energetic state of birds during migratory stopover. After landing in a random location within 10 km of the Gulf of Mexico, the "virtual" birds moved from cell to cell across the habitat map selecting the adjacent cell with the highest habitat value at each iteration of the model. After visiting each cell, the ESI of a bird was incremented to account for the amount of energy gained (due to foraging) and lost (due to energetic costs of foraging and movement) while occupying that cell.

Foraging costs were held constant for all habitat types, but the foraging gain accrued by birds as they moved across the landscape was determined by the habitat type of the cells the birds encountered. A bird's ESI was updated as it moved from habitat cell to habitat cell in the simulations. In productive habitats, migrants experienced a net energy gain (ESI gain > ESI cost). In poor habitats, migrants experienced a net energy loss (ESI gain < ESI cost). Foraging gains reflected our estimate of habitat quality based on field observations of the relative abundance of birds in these habitats. Four habitat categories were created from the original habitat types. Foraging gains equaled 0.1 in category 1 (poor) habitats, 0.25 in category 2 habitats, 0.8 in category 3 habitats, and 1.0 in category 4 (rich) habitats. Foraging costs were fixed at 0.5. The pattern of movement from cell to cell was determined by variation in habitat quality in adjacent cells. The model also incorporated a northward bias in movement to reflect the tendency for birds to orient northward during spring migration (Gauthreaux 1991). Birds moved from the current cell to one of the adjacent cell by choosing the cell with the highest value of the following expression: NBIAS*GAIN. NBIAS is a coefficient (range 0–1.00) representing the northward bias. NBIAS has the following values: 1.00 for the cell directly north (N) of the current cell, 0.75 for cells to the NW and NE, 0.50 for cells to W and E, 0.25 for cells to SW and SE, and 0.10 for the cell directly south (S). GAIN is the habitat-dependent foraging gain listed in the previous paragraph. Birds were not allowed to return to previously visited cells. In the individual-based model, a virtual bird began with an ESI of 10.0 and continued moving until it crossed one of two energy thresholds. If it gained enough energy (ESI ≥ 30.0), it left the study landscape on another long-range migratory movement. If its ESI dropped low enough (ESI < 2.0) because it failed to find productive habitats and lost energy, it ran out of energy and died. When an individual either migrated or died, the number of cells visited was recorded. In this way, the relative suitability of different landscapes could be examined by simulating a large number of individuals and keeping track of mortality and the number of cells visited before migration. Higher quality landscapes were characterized by low mortality and a lower numbers of cells visited by successful migrants.

RESULTS

Energetic Parameters

Table 1 summarizes spring data on arrival weight and condition collected from 1987–1992 on Horn Island and East Ship Island, Mississippi, for 14 common trans-Gulf migrants. The mean mass of "0" fat-class birds is close to the fat-free weights obtained in the laboratory (Dunning 1993). The span of annual mean weights measured in the field ranged from approximately fat-free levels, to weights indicating fat stores of about 10% body weight. These data provide reasonable estimates of the variability of energy stores to be expected among spring migrants arriving along the northern coast of the Gulf of Mexico following trans-Gulf migration.

Flight Performance Parameters

Applying these fat store estimates to the flight performance models developed by Pennycuick (1989) provides an estimate of the potential flight ranges of migrants after their arrival at coastal stopover sites (Table 1). Minimum range estimates, based on the range of mean annual arrival weights, indicate that in some years many birds are incapable of further migratory movement (flight ranges of tens of kilometers). Average arrival weights for the period 1987–1992 suggest ranges of tens to several hundred km for most species, while under the best of conditions ranges can exceeded 500 km. While observational evidence indicates that migration is concentrated during periods of favorable weather (Buskirk 1980, Gauthreaux 1991), prevailing winds will scale potential flight ranges up or down. For example, a 4 m/sec (14.4 km/hr) head wind reduced these range estimates by approximately 50%, while a 4 m/sec tail wind increased

TABLE 1. ARRIVAL WEIGHTS AND POTENTIAL FLIGHT RANGES OF COMMON TRANS-GULF MIGRANTS

Species[a]	N (1987–1992)	Wing span[b] (m)	Wet weight[c] (g)	Fat free mass[c] (g)	Mean mass "0" fat-class birds (g)	Mean arrival mass all birds (g)	Range of mean annual arrival weights[d] (g)	Potential range[e] still air (km)			Potential range (mean fat levels) with wind (km)	
								Minimum	Mean	Maximum	4 m/sec head wind	4 m/sec tail wind
RTHU	31	0.1100	3.64	2.62	2.88	2.9	2.87–2.96	0	21	83	10	32
COYE	285	0.1725	10.78	8.36	8.86	9.14	7.93–9.61	0	114	298	63	165
BGGN	44	0.1830	6.00	4.67	5.69	6.16	5.45–8.13	0	310	1390	165	431
NOPA	82	0.1885	7.60	5.93	6.69	6.97	6.20–7.46	0	161	429	88	234
WEVI	1554	0.1940	10.04	8.29	10.04	10.65	10.27–10.99	89	231	353	124	338
SWWA	42	0.1960	18.90	14.70	13.28	13.59	12.77–16.88	0	88	915	52	121
PROW	176	0.1990	13.94	11.01	11.37	11.94	10.72–13.18	0	192	579	104	270
HOWA	680	0.2035	10.29	8.20	9.11	9.42	8.86–10.43	0	135	547	75	195
ACFL	110	0.2050	12.90	10.03	11.38	11.75	11.43–11.93	18	128	188	70	181
INBU	825	0.2100	15.45	12.34	12.08	13.16	12.07–14.16	0	346	641	157	406
REVI	2462	0.2400	18.88	14.59	14.34	15.6	14.38–17.56	12	363	874	194	534
SUTA	370	0.2860	36.50	24.73	25.95	26.94	24.45–29.04	0	169	508	99	232
GCFL	56	0.2870	33.50	26.06	32.11	32.6	29.79–33.49	0	67	185	42	90
WOTH	416	0.3270	56.92	42.21	39.03	40.6	39.15–44.98	14	122	664	76	168

[a] Species Codes: RTHU Ruby-throated Hummingbird, *Archilochus colubris,* COYE Common Yellowthroat, *Geothlypis trichas,* BGGN Black-throated Green Warbler, *Dendroica virens,* NOPA Northern Parula, *Parula americana,* WEVI White-eyed Vireo, *Vireo griseus,* SWWA Swainson's Warbler, *Limnothlypis swainsonii,* PROW Prothonotary Warbler, *Protonotaria citrea,* HOWA Hooded Warbler, *Wilsonia citrina,* ACFL Acadian Flycatcher, *Empidnonax virescens,* INBU Indigo Bunting, *Passerina cyanea,* REVI Red-eyed Vireo, *Vireo olivaceus,* SUTA Summer Tanager, *Piranga rubra,* GCFL Great-crested Flycatcher, *Myiarchus crinitus,* WOTH Wood Thrush, *Hylocichla mustelina.*

[b] Source: HOWA, NOPA, COYE, REVI, WOTH, RTHU from R. Mulvihill, Powder Mill Nature Reserve; WEVI from M. Woodrey, MS Nat. Hist. Mus.; BGGN, GCFL, PROW, ACFL, SWWA from D. Evered, FL State Univ.; INBU from Horn Island 1994 (F. Moore, unpubl. data).

[c] Source: HOWA, REVI, WEVI, INBU, COYE, NOPA, PROW, SUTA, WOTH, RTHU from E.P Odum in Dunning (1993); BGGN, GCFL, YTVI, ACFL, SWWA estimated as % wet weight (77.79%) reported in Dunning (1993).

[d] Source: Spring 1987-1992, East Ship Island and Horn Island, MS (F. Moore, unpubl. data).

[e] Source: Pennycuick 1989, Prog_1.bas for analysis of powered flight. Estimates assume air at altitude of 350 m above sea level and "0" fat mass for empty body mass.

TABLE 2. HABITAT ASSOCIATIONS[a] OF COMMON TRANS-GULF MIGRANTS IN THE COASTAL ZONE OF MISSISSIPPI

	1992		1993		
Species[b]	Bottom-land	Pine	Bottom-land	Pine with under-story	Pine without under-story
HOWA	249	18	161	126	0
REVI	230	16	211	13	0
WEVI	203	70	77	52	0
BGGN	82	2	117	21	1
GCFL	47	6	66	22	6
INBU	15	63	11	4	24
COYE	16	31	6	32	69
NOPA	47	8	39	0	0
YTVI	42	9	31	4	4
PROW	62	8	16	0	0
ACFL	45	0	26	1	0
SUTA	21	15	18	28	7
WOTH	32	1	15	3	0
RTHU	17	2	15	6	2
SWWA	14	1	3	0	0
Total individuals	1122	250	812	312	113
%	82	18	66	25	9
Total species	43	26	40	30	16

[a] 1992 = 9 sites × 7 replicates = 63 1-km strip transect censuses/habitat (2 habitat types/site) ($F = 7.09$, $P < 0.01$); 1993 = 9 sites × 6 replicates = 54 1-km strip transect censuses/habitat (3 habitat types/site) ($F = 4.87$, $P < 0.01$). Numbers represent total number of individuals recorded in each habitat type.

[b] See Table 1 for species codes.

them by a similar amount (Table 1). The effects of head and tail winds can be used in this model to simulate the variability in weather conditions encountered by migrants.

HABITAT PARAMETERS

Censuses at mainland and barrier island stopover sites indicate that birds select habitats nonrandomly during migration. We have found that, although scrub/shrub and forest habitats accounted for 20% of the available habitat, they were associated with over 70% of the migrants observed in censuses on Horn Island, Mississippi (Moore et al. 1990). Censuses conducted during the spring of 1992 and 1993 at adjacent mainland sites showed that the number of individuals and total number of species detected was considerably greater within riparian bottomlands and pine forests with a well developed shrub understory than in other habitats. Approximately 80% of all detections were in these two habitat types (Table 2).

We assume that the differences in habitat preference that we have observed in the field reflect real differences in habitat quality. However, our understanding of the quantitative differences between habitats is still very limited. Some evidence is available from measurements of migrant turnover rates and estimates of prey availability made at stopover sites.

We have found that birds without fat stores are more likely to be recaptured at stopover sites (Kuenzi et al. 1991, Moore and Simons 1992a), suggesting that birds with sufficient energy stores resume migration sooner or select better habitats. We have also documented differences in recapture rates at different stopover habitats. For example, 20.7% ($N = 8,392$ total captures, 1988–1991) of the birds stopping at Peveto Beach in southwest Louisiana stay one or more days and are recaptured versus 8.9% ($N = 12,080$ total captures, 1987–1991) at East Ship Island, Mississippi ($P < 0.001$). Again, we interpret this difference to be a reflection of habitat quality. Rates of mass gain during stopover are generally higher at the Louisiana site (Fig. 2a), which is consistent with measurements of higher insect prey densities at that site (Fig. 2b). Until we understand more fully the factors that determine the quality of migratory bird stopover habitats, we will be limited to grouping habitats into fairly coarse categories of habitat quality. Nevertheless, habitat groupings that rank habitats according to their suitability for passage migrants are useful for exploring the effect of landscape-level patterns of habitat availability.

ANALYSIS OF SPATIAL PATTERN AND STOPOVER HABITAT SUITABILITY

In an initial attempt to explore how variability in habitat quality might affect migrants that depend on coastal stopover habitats, we reduced the 18 cover types of our original landcover map to four habitat categories. These categories reflected the relative abundance of migrants in coastal habitats based on our experience and the results of our field censuses (Table 3). These ranged from category 1 habitats (urban, industrial, open water, and beach habitats), which were classified as unsuitable, to category 4 habitats (wetland-forested and deciduous bottomland forest), which we believed to represent the richest stopover habitat types. We then subdivided the coastline into five study areas of approximately 1200 km^2 each and ranked the areas according to their average habitat rank. Ranks reflected the average habitat score calculated from the reclassified cells within each study area (Fig. 3). Area 2 had the lowest habitat rank followed by areas 3, 4, 1 and area 5 with the highest habitat rank.

Several spatial indices were calculated for areas 1 and 2 as an example of how measures such as contagion can be applied to stopover habitats (Table 4). In this comparison, the contagion indices are similar. That is, the probability that two adjacent 28.5 m x 28.5 m cells will be of the same habitat type is similar in both areas. On the other hand, the juxtaposition of cells of dif-

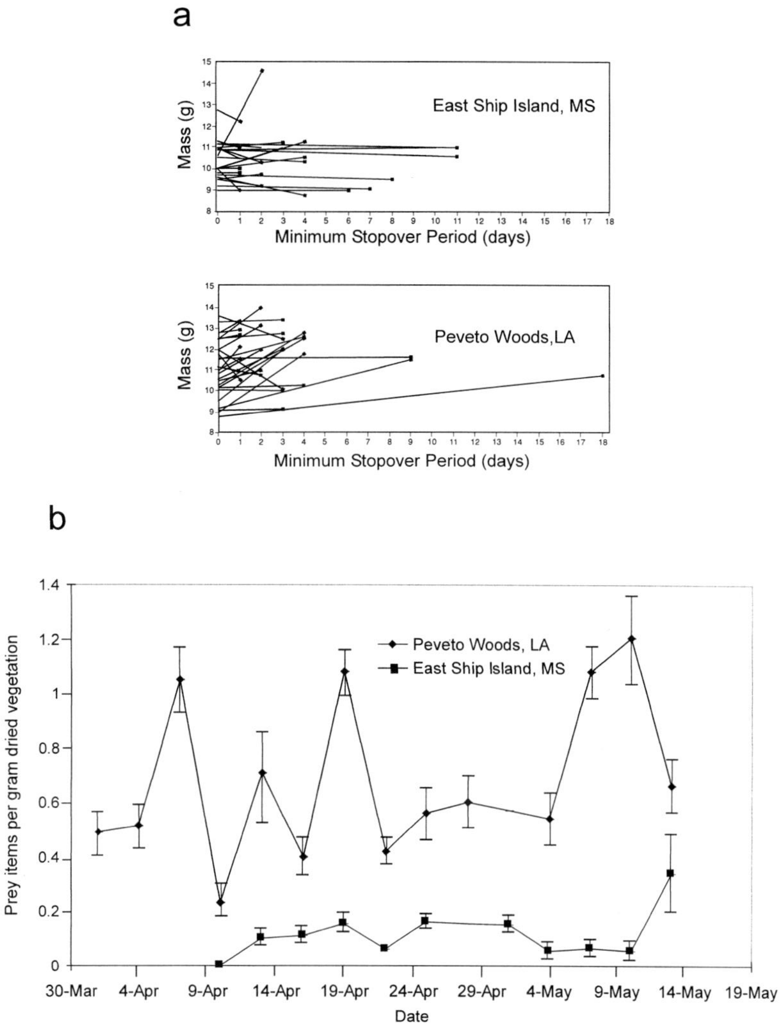

FIGURE 2. Evidence of variability in stopover habitat quality. (a) Weight trajectories (first and last capture) of individual White-eyed Vireos at stopover sites in southwest Louisiana (N = 33) are consistently higher than those on the Mississippi barrier islands (N = 30). (b) Abundance of prey for foliage gleaning birds is consistently higher ($P < 0.05$, Student's t-test) at the Louisiana stopover site. See Kuenzi et al. (1991) for sampling methods.

ferent habitat types, an edge index, suggests an important difference between the two areas. The probability that cells of low quality (category 1 or 2) habitat will be adjacent to cells of high quality habitat (category 4) is significantly greater in area 1 than in area 2. These transition probabilities may not be important to migrants that arrive along the coast with significant energy stores (i.e., potential ranges of hundreds of km), but they may be very significant to birds with depleted stores and limited ability to search for suitable stopover habitats.

The window analysis allowed us to quantify the variation in landscape-level foraging opportunities experienced by arriving migrants. Within the same landscape, there are likely to be rich as well as poor areas, but an individual bird can only use a small portion of the available habitat due to ecological, morphological, and energetic constraints. Figure 4a illustrates two windows randomly placed in Study Area 5. In the analysis, the size of the window was allowed to vary to simulate the variability in the energetic state of birds arriving in stopover habitats following trans-Gulf flights. For the purpose of this analysis, the window radius simulated birds arriving with effective ranges of from 1–30 km, the lower range of mobility estimated from field and flight performance data.

The technique allowed us to analyze how the

TABLE 3. HABITAT CATEGORIES USED IN SPATIAL MODELS

	Category 1	Category 2	Category 3	Category 4
	unclassified	wetland/emergent marsh	wetland/mixed scrub-shrub	wetland/forested/deciduous
	water	residential	upland/woody/evergreen/scrub-shrub	wetland/forested/deciduous/bottom-land forest
	excavated soil	upland/woody/pine forest	upland/woody/mixed/shrub-scrub	
	beach/sand	upland/herbaceous cropland	upland/woody/mixed forest	
	wetland/sand bar	upland/orchards		
	commercial			
	transportation			
	industrial			
% Total landscape	7.5	53.6	31.4	7.5

Note: Category 1 habitats were assumed to represent the poorest habitats for migrants, category 4 habitats were assumed to represent the best habitats for migrants.

energetic state of arriving birds affected their ability to use available habitats. Figure 5a depicts how increasing the window radius (simulating arriving birds with improving energetic states) affects the mean habitat rank (quality) of the habitats available to migrants. While the lack of a trend may reflect the relatively homogeneous nature of the habitats at this scale, habitat variability appears to decline as the window radius increases, suggesting that habitat suitability thresholds may exist for birds during stopover. This specific result could simply be a sampling artifact, but a similar analysis across a range of landscapes may reveal patterns that improve our understanding of how energetic status and the degree of habitat specialization interact to shape the stopover ecology of migrants. Certainly, the variability in habitat quality in a landscape might be just as important to some migrants as average habitat conditions.

We also examined variability in habitat quality among our study landscapes. Figure 5b shows the mean habitat rank of 50 10-km radius windows randomly placed in each of the five study areas. The richest study area (area 5) showed less variability than the poorer habitats (areas 2, 3, and 4). Again, the biological significance of these patterns is probably a function of the scale at which birds are sampling stopover habitats. For example, in spite of the fact that area 4 (Fig. 3) contains a corridor of rich deciduous bottomland forest, birds arriving in the area with an effective range of 10 km will on average encounter habitats that are of lower quality than the area as a whole (Fig. 5b). Rescaling the analysis, by increasing the effective range to simulate birds arriving with more fat, or reducing the effective range to simulate the effects of headwinds, would undoubtedly alter the rankings of the sites.

Individual-based models provide another tool to evaluate how the spatial pattern and quality of stopover habitats may affect trans-Gulf migrants. Several examples will illustrate how we have applied individual-based models to these questions. The basic premise of the model is that on rich landscapes few individuals should die, and the number of cells visited should be low, while on poor landscapes more individuals will die, and the number of cells visited by successful migrants is expected to increase. Figure 4b illustrates the movement of two "virtual" birds placed randomly within a study landscape. Note that the birds tend to track the richer (darker) habitat types. We might predict that the effects of landscape quality and arrival condition on the movement and survival of birds will not be strictly additive. For example the model can be used to examine

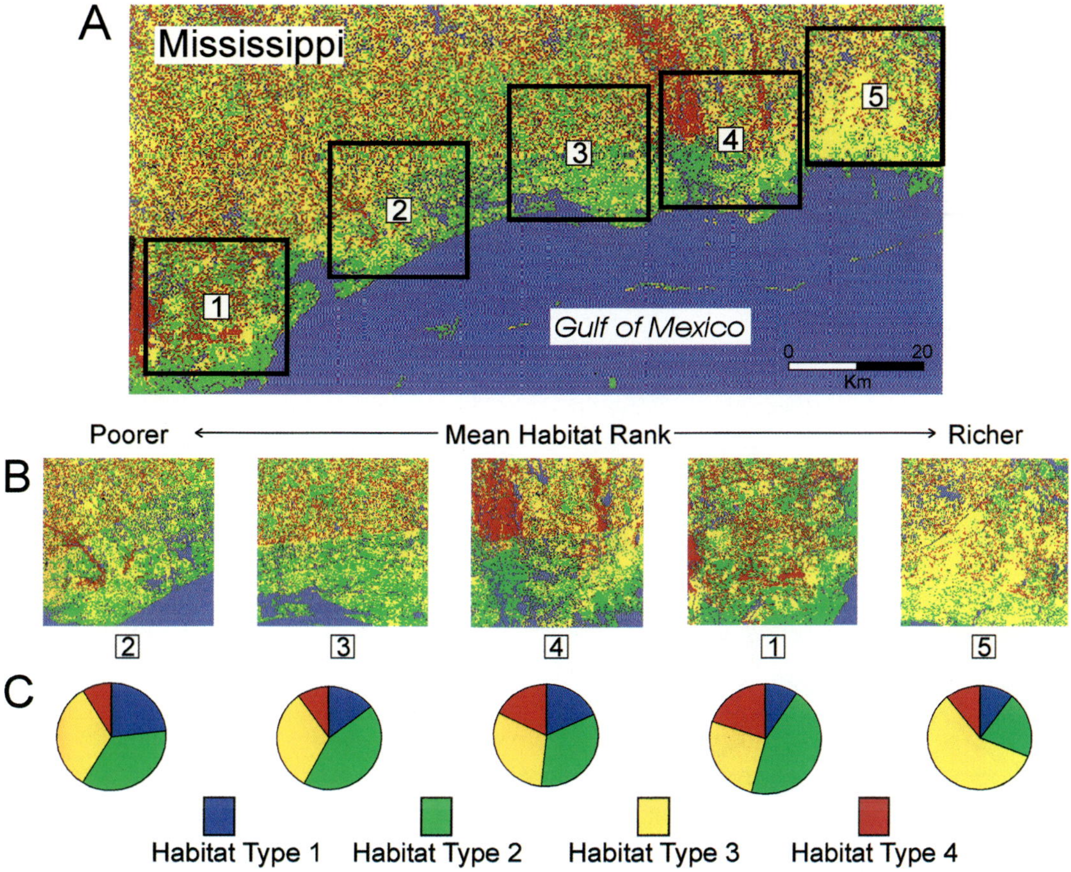

FIGURE 3. Composition of coastal habitats. Five study areas were selected and classified according to the categories described in Table 3. Mean habitat ranks were calculated for each study area based on the abundance of habitats in each of the four categories. Mean habitat ranks for the individual study areas were: Area 2 (2.27), Area 3 (2.38), Area 4 (2.47), Area 1 (2.56), Area 5 (2.69).

TABLE 4. SPATIAL INDICES FOR AREAS 1 AND 2

Index	Area 1	Area 2
Contagion[a]	0.389	0.388
Edge Index[b]		
1 and 2	27484	49007
1 and 3	26518	65183
1 and 4	10717	6211
2 and 3	147589	194881
2 and 4	137474	47672
3 and 4	81223	61347

[a] The probability that two adjacent cells will be of the same habitat type.
[b] A measure of the contrast between adjacent cells, e.g., the probability that a high quality habitat cell will be adjacent to a low quality cell.

whether birds that arrive with very low energy reserves experience disproportionately greater rates of mortality and slower rates of energy gain and if so, how those rates vary with changes in average habitat quality.

Simulations of 200 hypothetical individuals showed that both habitat quality and the arrival energy state index (ESI) affected the percentage of birds that survived to continue migrating (Fig. 6). It appeared that a bird's energetic state upon arrival was most significant in landscapes of intermediate habitat quality. In very rich (high habitat rank) or very poor (low habitat rank) landscapes, arrival ESI was not well correlated with survival. Landscape suitability, as measured by habitat rank, affected both the mean and variance of the number of cells visited by simulated migrants (Fig. 7). These trends suggest that the relationship between these factors

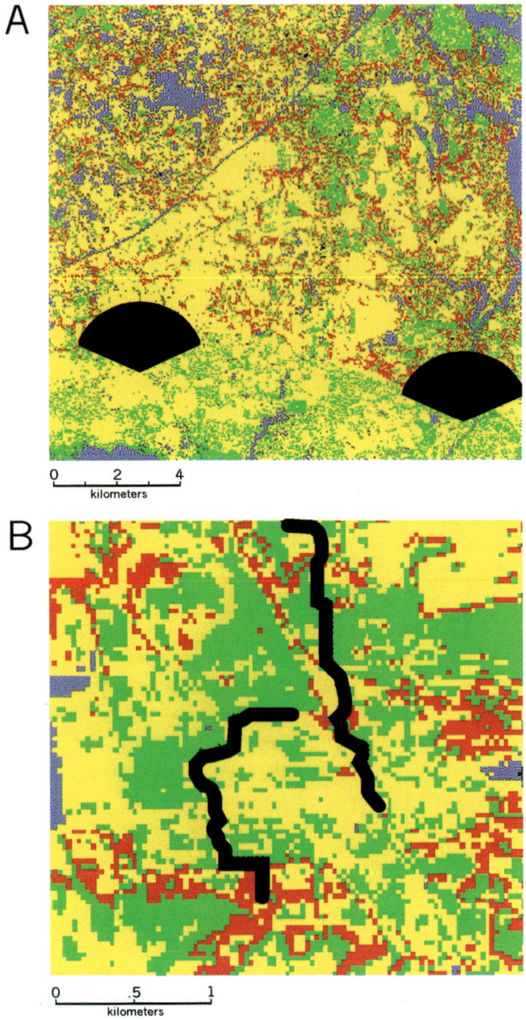

FIGURE 4. Window analysis. (a) Random projection of two windows over study area 5. Shape of window reflects migrant's tendency to move northward during spring migration. Size of window represents energetic state upon arrival. Cell size 90m x 90m. (b) Individual-based model. Movement of two "virtual" migrants placed randomly in a study landscape. Birds tend to track richer (darker) habitat types.

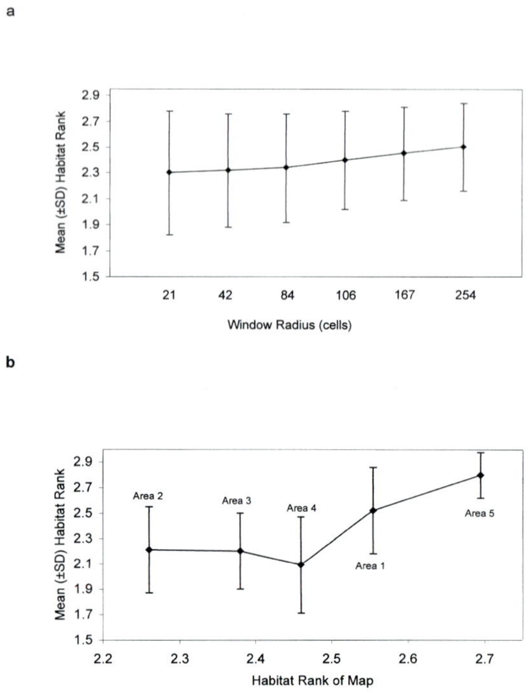

FIGURE 5. Window analysis. (a) Relationship of window size (radius from 1–30 km) to mean habitat rank (N = 50 windows at each radius). (b) Mean habitat rank of 50 10-km windows versus the habitat rank of the entire study area map.

is probably not linear, and that the variance in the number of cells visited decreases in richer habitats. As we might expect, the arrival ESI is inversely related to the mean number of cells visited by migrants that survive to continue migration (Fig. 8).

An analysis of variance tested for the effects of mean habitat rank (MAP) and the arrival energetic state (ESI) on the number of cells visited by individuals that survived to migrate. The model used was: Cells visited = MAP + ESI + MAP x ESI. This analysis showed that both the study landscape (Fig. 7; $F = 226.71$, $df = 4$, $P < 0.001$) and the energetic state of arriving birds (Fig. 8; $F = 35.69$, $df = 3$, $P < 0.001$) significantly affected the number of cells that migrant birds visited. Moreover, because the interaction term is significant ($F = 6.04$, $df = 12$, $P < 0.001$) we know that the effects of landscape and ESI are not strictly additive. Figure 9 provides

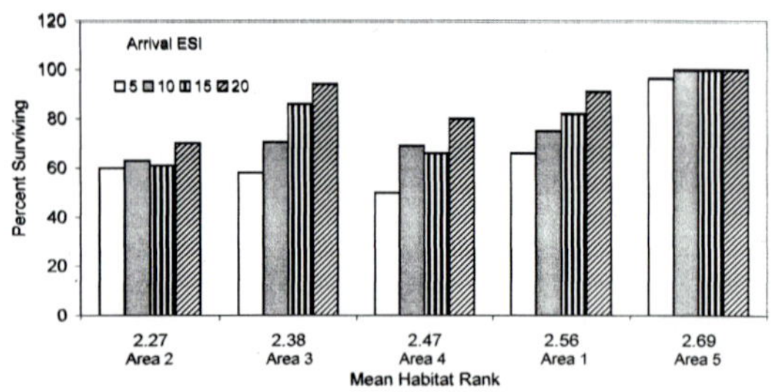

FIGURE 6. Effect of arrival energetic state (ESI) and habitat rank on the percentage of individuals surviving in the individual-based model.

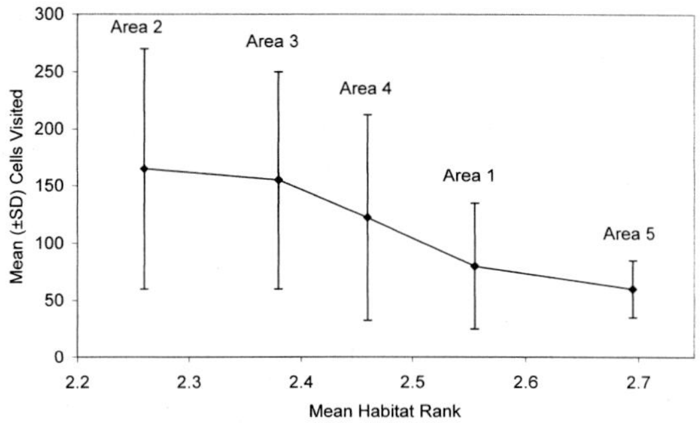

FIGURE 7. Relationship between mean habitat rank of the study area and the mean number of cells visited by 200 "virtual" migrants in the individual-based model.

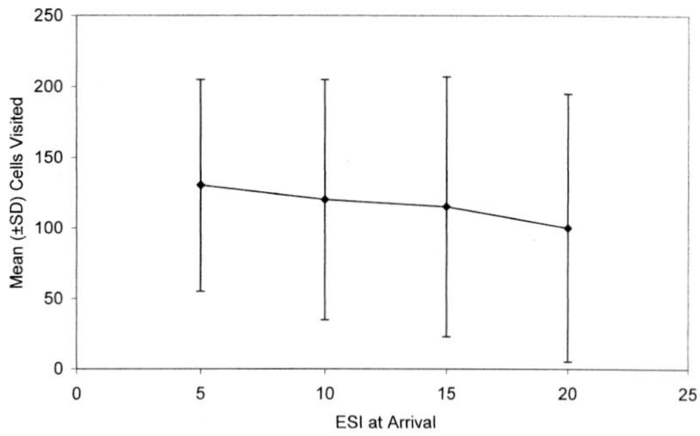

FIGURE 8. Influence of arrival energetic state (ESI) on the mean number of cells visited by "virtual" migrants that survived to migrate.

evidence that the effect of arrival ESI was greater in the richer landscapes (especially areas 1 and 5). ESI was not a good predictor of the number of cells visited on the poorer landscapes (areas 2 and 3).

DISCUSSION

Spatial models allow us to explore the interplay of organisms and the landscapes they occupy, in particular the relationship between the ecology and behavior of individual species and the spatial variability of the habitats they occupy. We believe that the quality and spatial pattern of habitats, and the energetic status of birds when they arrive at stopover sites impose important constraints on the likelihood that individual birds will migrate successfully.

Techniques such as window analysis allow us to examine how variations in the energetic state of arriving birds and local weather conditions determine the scale at which birds experience stopover landscapes. Individual-based models, while having more assumptions, allow us to conduct a sensitivity analysis of the relative importance of physiological and ecological constraints, and they suggest new hypotheses to test with field data. For example, by projecting current trends in habitat conversion into the future, we can explore the potential impact on species with differing habitat requirements and flight ranges, or how the interplay of habitat patchiness and arrival energetic state affect the likelihood of a successful migration. Behavioral characteristics of migrants, such as territoriality (Rappole and Warner 1976) and ecological plasticity (Greenberg 1990) can also be incorporated into these models. Such refinements will require better information on the behavioral ecology and habitat requirements of individual species, and the status and trends of the habitats they occupy. As Moore and Aborn (*this volume*) have shown, radio telemetry holds tremendous promise for improving our knowledge of the ecology of migrants at stopover sites. Larger scale studies, while logistically challenging, would also seem well warranted.

Information of this type will be particularly important as landscapes become increasingly modified by human activity. Recent projections indicate that coastal communities surrounding the Gulf of Mexico are likely to experience significant population growth over the next 15–20 years (Fig. 10). If patterns of habitat loss elsewhere are a guide, we can predict that the coastal deciduous and riparian bottomland habitats that are clearly important to migrants will be lost at a disproportionately high rate. We feel that spatial models integrating information about the ecological requirements of migrants and the spatial patterns of stopover habitat will be essential in helping to set research and conservation priorities in the future.

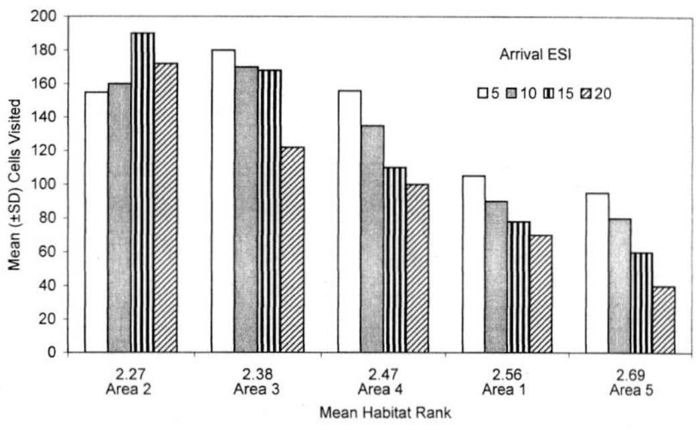

FIGURE 9. Interaction of arrival energetic state (ESI) and habitat rank of the study area on the mean number of cells visited by "virtual" migrants that survived to migrate.

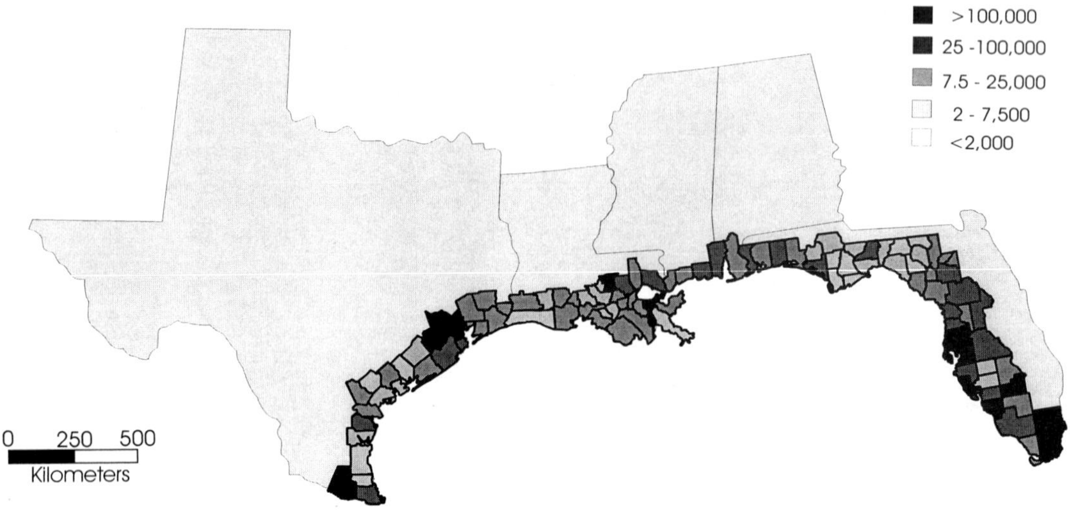

FIGURE 10. Projected population growth by county along the northern Gulf coast 1988–2010 (Culliton et al. 1990).

ACKNOWLEDGMENTS

We thank D. Evered, R. Mulvihill, and M. Woodrey for unpublished wing-span data. P. O'Neil, U.S. Geological Survey, produced the original Landsat classification. R. Caldow, J. Clark, B. Danielson, J. Goss-Custard, and S. Mabey provided valuable comments on the manuscript. Funding for this research was provided by the U.S. Fish and Wildlife Service, the National Park Service, the National Biological Service, the U.S. Forest Service, the National Science Foundation, and Oak Ridge National Laboratory.

Studies in Avian Biology No. 20:15–33, 2000.

HABITAT USE BY LANDBIRDS ALONG NEARCTIC-NEOTROPICAL MIGRATION ROUTES: IMPLICATIONS FOR CONSERVATION OF STOPOVER HABITATS

DANIEL R. PETIT

Abstract. Most wildlife management and conservation plans are based upon patterns of habitat use by focal species. Lack of information on habitat use by birds during migration has prevented development of comprehensive strategies for their protection along migration routes, including identification of high priority habitat types and specific sites critical to long-term persistence of those species. In this review, published information about habitat associations of long-distance migrants along nearctic-neotropical migration routes was used to address several relevant questions about the patterns, proximate and ultimate causes, and management implications of habitat use during the migration period (primarily in North America). Most species used a restricted set of habitats from those available. In general, however, species were more variable in their use of habitats during migration than during the breeding season, and they exhibited substantial variation in use of habitats at different locations along migration routes and between spring and autumn migration periods. General patterns of habitat use by species during migration corresponded most closely to patterns of habitat use during the breeding season rather than to measures of the types or abundance of food found within habitat types, competition from other species, or presence of predators during migration. These preliminary results suggest that specific guidelines developed for conservation of migratory species during the breeding season will be useful for their management during migration periods as well. In addition, large tracts of structurally diverse forests, natural representation and distribution of habitats within landscapes, and sites adjacent to geographic barriers (large bodies of water, mountain ranges) should be of high priority for conservation of the stopover habitats of migratory birds.

Key Words: conservation priority, habitat use, migration, nearctic-neotropical migrants, North America, stopover habitat.

> "Where do the birds go each fall that have nested in our dooryards and frequented the neighboring woods, hills, and marshes? Will the same ones return again to their former haunts next spring? What dangers do they face on their round-trip flight and in their winter homes? These and other questions puzzle the minds of many who are interested in the feathered species... Lack of information on the subject may mean the loss of an important resource by unconsciously letting it slip from us. Ignorance of the facts may be responsible for inadequate legal protection for such species as may urgently need it. More general knowledge on the subject will aid in the perpetuation of the various migrants, the seasonal habitats of some of which are in grave danger from man's utilization, sometimes unwisely, of the marsh, water, and other areas they formerly frequented."—Frederick C. Lincoln, *The migration of North American birds (1935)*

The connection between environmental problems and health of some bird populations in North America was first widely recognized during the 1960s (Carson 1962), but nearly three decades passed before the extent of those problems was fully realized for migratory birds as a group (Robbins et al. 1986, 1989b). During that period, avian ecologists interested in conservation and management of long-distance migratory land birds worked along parallel tracks during the breeding season in temperate North America and during the overwintering period at tropical latitudes (see Keast and Morton 1980, Hagan and Johnston 1992). Habitat loss and fragmentation were identified as the most pressing avian conservation problems in both areas (e.g., Aldrich and Robbins 1970, Forman et al. 1976, Morse 1980b, Whitcomb et al. 1981, Lynch and Whigham 1984, Hutto 1988).

Long-distance nearctic-neotropical migrants are those species that breed in temperate North America and overwinter at tropical latitudes. The annual cycle of most species entails spending 3–4 months at breeding sites, 5–6 months at overwintering areas, and the remaining 2–4 months along migratory routes (Keast and Morton 1980). However, despite the relatively greater risks to birds travelling several thousand kilometers along migratory routes, inadequate attention has been devoted to understanding the habitat requirements, behavioral ecology, and energetic constraints of birds during migration. Hence, the level of scientific investigation during migratory periods has not been commensurate with the probable role these periods play in the population dynamics of nearctic-neotropical migrants (Sprunt 1975, Gauthreaux 1979).

Only in the past few years has attention been

given to conservation of landbirds along migratory pathways in the Western Hemisphere (Moore et al. 1993). However, basic knowledge of the types of habitats used by species at stopover sites has remained elusive. Documentation of the patterns of habitat use, as well as understanding the proximate and ultimate bases for that behavior, are fundamental to effective conservation plans since many conservation and management actions are directed at habitats and only indirectly at species.

I address several questions of habitat use that are significant to nearctic-neotropical migratory bird ecology and conservation: (1) Do migrating birds exhibit nonrandom use of habitat types? (2) Are certain habitat types or vegetative characteristics consistently related to use by migrating birds? (3) Do species show consistent use of habitat types at different locations along migratory routes? (4) Are patterns of habitat use consistent between spring and autumn migratory periods? (5) How does habitat use during migration compare with that during winter and breeding periods? (6) What are the ecological correlates of habitat use along migration routes? (7) Are guidelines for management of species during the breeding season in North America appropriate for migration periods as well? Evaluation of these questions, which complements the recent reviews by Moore and co-workers (Moore and Simons 1992a; Moore et al. 1993, 1995), is intended to provide direction for identifying and managing migratory stopover habitats and for guiding future research efforts.

DO MIGRATING BIRDS EXHIBIT NONRANDOM USE OF HABITAT TYPES?

Migratory birds are not distributed haphazardly among habitats during either the breeding (Hamel 1992) or wintering (Petit et al. 1993) periods, so nonrandom habitat use by migrating birds also would be expected. Results from the few systematic studies that have examined this question during migration indicate that populations of most species are not distributed equitably across major habitat types (Parnell 1969, Mason 1979, Hutto 1985a, Moore et al. 1990, Mabey et al. 1993). For example, the distribution of most species across habitats is highly skewed, such that habitat breadth (see Levins 1968) of individual species rarely reaches 50% of the maximum possible (Fig. 1, shaded bars; a mean of 40%, for example, indicates that the breadth of distribution of individuals across available habitats averaged only 40% of the value were individuals equally distributed across habitat types), and most species typically are not even detected in one-third of the available habitats (Fig. 1, diagonal bars; a mean of 65%, for example, indicates that the "average" species was detected in 65% of all habitats surveyed). Thus, migrating birds exhibit selective use (defined as deviation of use from availability) of some habitats over others.

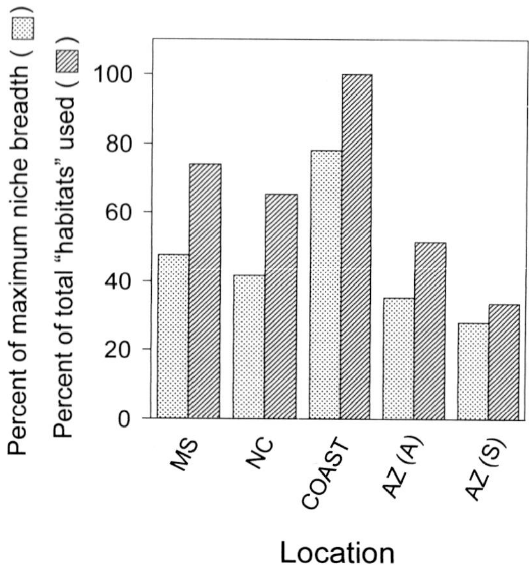

FIGURE 1. Examples of the overall distribution of migratory birds across available habitats in Mississippi (MS; Moore et al. 1990), North Carolina (NC; Parnell 1969), the mid-Atlantic coast (COAST; Mabey et al. 1993), and Arizona in autumn (AZ[A]) and spring (AZ[S]; Hutto 1985a). Percent of maximum niche breadth was derived by calculating the niche breadth (Levins 1968) of each species as a percentage of the maximum value possible, and then averaging over all species. Percent of maximum habitats used was calculated in a similar fashion, except that niche breadth was replaced by the percentage of all habitats occupied by each species, and then averaged over all species. (Measures are conservative estimates of the distribution of birds across habitats because most studies included only relatively abundant species and omitted uncommon and rare species that most likely had more restricted distributions.)

Habitat selectivity varies widely among species, however. For example, in the lower Piedmont of North Carolina, Parnell (1969) found that Yellow-rumped (*Dendroica coronata*) and Black-and-white (*Mniotilta varia*) warblers were broadly distributed, while Yellow (*D. petechia*) and Prothonotary (*Protonotaria citrea*) warblers were detected in only two of seven habitat types. Likewise, Golden-crowned Kinglets (*Regulus satrapa*) migrating through southeastern Arizona were restricted to high elevation pine-fir forests, whereas Ruby-crowned Kinglets (*R. calendula*) moving through the same region were detected in a wide variety of habitat types (Hutto 1985a). Other studies have documented similar variation

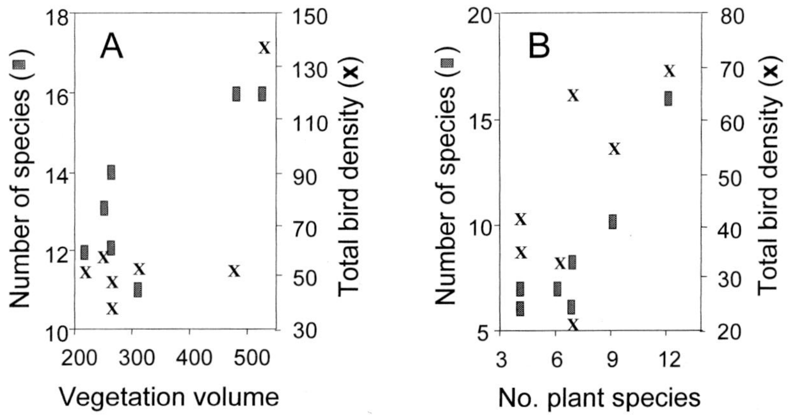

FIGURE 2. Relationship between measures of bird community composition (species richness represented by squares, total density of birds represented by crosses) and vegetative characteristics (volume of vegetation and woody plant species richness) during (a) autumn and (b) spring migrations in southeastern Arizona (Hutto 1985a).

in the breadth of species' habitat use during migration.

In summary, most migratory species exhibit selective use of locally-available habitats during migration, much as they do during other seasons. Many species concentrate locally in up to three habitat types (e.g., Hutto 1985a, Moore et al. 1990), with fewer individuals distributed among remaining habitats. However, as discussed above (and below), those apparent local preferences are both geographically and temporally flexible. This raises the question of whether certain major habitat types, or specific vegetative characteristics common to several habitats, are favored by migrating birds.

ARE CERTAIN HABITAT TYPES OR VEGETATIVE CHARACTERISTICS CONSISTENTLY RELATED TO USE BY MIGRANTS?

Because human societal values are not consistent with protecting all areas and habitat types necessary to sustain healthy populations of migratory birds, a serious dilemma is faced by those developing plans for the conservation of migration stopover sites: Which habitats are most critical to protect?

MacArthur and MacArthur (1961) and others (e.g., Willson 1974, Terborgh 1977, Beedy 1981) have empirically demonstrated the intuitive relationship between structural complexity of habitats and bird species diversity in both temperate and tropical areas. This relationship, however, breaks down when examining species diversity across habitats of relatively similar structure and plant species composition (e.g., Roth 1976, Szaro and Balda 1979, Erdelson 1984, Petit et al. 1985). Although the above paradigm has important ramifications for conservation of priority habitats or areas, it has not been addressed specifically for migratory birds occupying stopover habitats.

Several studies provide general support for the relationship between foliage complexity and bird species richness and abundance during migration. Moore et al. (1990) found that migrants arriving at the Gulf coast of Mississippi during spring were most diverse and abundant in pine forests and in 5-m-tall shrub habitats, and were least common in dunes and marshes. Sykes (1986) observed a similar pattern on North Carolina barrier islands during autumn migration. Blake (1984) showed that species richness and abundance of migrating birds were correlated with vegetation height and density across three plots in southern Nevada; that relationship, however, may have been confounded by elevational factors. Both Martin and Vohs (1978) and Yahner (1983) found that abundance and diversity of transient birds moving through the Great Plains were positively associated with measures of foliage diversity. Beaver (1988) suggested that the increased autumn bird use of irrigated old fields, compared to nonirrigated fields, may have been due to greater vegetative biomass (or arthropod abundance) on the former sites. Hutto (1985a) has gathered perhaps the most detailed data to address this hypothesized bird-habitat association. For birds migrating through Arizona, a general positive relationship was observed between vegetation characteristics (e.g., volume of vegetation, number of woody plant species) and bird species richness and density during both spring and autumn across seven sites (Fig. 2). In that study, all 10 of the correlation coefficients between bird community attributes and vegetation characteristics were positive during autumn, and 9 of 10 were positive during spring. In both autumn and spring, birds migrating through old-growth hammocks in Florida appeared to be attracted to areas with heterogeneous and complex vegetation—forest edges, natural gaps, and areas with dense understory (Noss 1991).

Several studies, however, have found little evidence of a relationship between foliage complexity and measures of bird use. Spring migrants travelling through North Carolina (Parnell 1969) were slightly more abundant in low thickets ($\bar{x}$ = 14.1 ± 1.1 SD birds/hr) than in taller forests (11.7 ± 2.0; Mann-Whitney U-test, Z = 1.36, P = 0.17), although that nonsignificant trend was reversed when species richness was examined (thickets, $\bar{x}$ = 7.5 ± 4.9 SD species; forests, 14.4 ± 3.6; Z = −1.36, P = 0.17). Along the Delmarva and Cape May peninsulas of the Atlantic coast, no consistent relationships were obvious between bird species richness or abundance and the structural complexity of 17 plant community types (Mabey et al. 1993). Likewise, data in Weisbrod et al. (1993) suggest only a weak relationship between birds and habitat complexity. This latter data set, however, was based upon mist-netting and, therefore, probably was biased against taller vegetation types. In Arizona, numbers of both fall and spring migratory species passing through ponderosa pine (*Pinus ponderosa*) forests were lowest on sites with a high density of overstory trees and greatest on plots with many shrubs and saplings (Blake 1982). In contrast, total abundance of spring migrants in Blake's study was inversely related to understory density, while abundance of autumn migrants showed no relationship with either understory or overstory. In wooded riparian corridors of southeastern Arizona, Skagen et al. (1998) found no significant relationship between foliage density and either species richness or abundance of migrants.

In summary, at least as many (and often more) species and individuals are typically found in structurally diverse habitats compared to less diverse sites. However, the lack of a consistent relationship between bird community and vegetative characteristics probably results from the cumulative effects of species-specific responses to habitat structure. That is, each species responds to a unique set of environmental stimuli, such that divergent responses by the different species are likely to obscure a definitive pattern of habitat use by the bird community as a whole.

The meager information on avian use of vegetation types during migration, and the dynamic nature of plant communities across geographic regions, makes it difficult, and indeed probably academic, to identify specific plant communities most important as stopover habitat (but see below). Rather, examination of the suite of habitats on a local or sub-regional level may be an appropriate scale at which to identify habitats most beneficial to migrants as a group.

In general, taller, more structurally diverse vegetation types within an area appear to support greater numbers of migrating birds than do habitats of lower stature and complexity. Clearly, those structurally complex habitats will not be adequate for all migratory species, but if a conservation goal is to protect those areas used most frequently by migrating birds, relatively tall, structurally diverse habitats may best serve that purpose. The plasticity in habitat use exhibited by most species during migration (see above) suggests that many species are able to effectively use the food resources and cover afforded by structurally complex habitats. Additional research is needed on this topic, however, as simple presence may not reflect the quality of a site, but rather "forced" selection driven by low energy stores after overnight flights (Hutto 1985b, Moore and Kerlinger 1987, Moore and Simons 1992a).

DO SPECIES SHOW CONSISTENT USE OF HABITAT TYPES AT DIFFERENT LOCATIONS ALONG MIGRATION ROUTES?

Many species show substantial geographic variation in habitat use, even among those studies where similar habitats were examined. For example, in a comparison of nine species of wood-warblers migrating through both the Piedmont of North Carolina (Parnell 1969) and along coastal areas several hundred kilometers to the north (Mabey et al. 1993), average within-species overlap (Colwell and Futuyma 1971) in habitat use between the two areas was only 63% (SE = 5.3, range = 38–84%). Yellow Warblers migrating through eastern coastal areas (Mabey et al. 1993), North Carolina (Parnell 1969), and Wisconsin (Weisbrod et al. 1993) nearly always (93–100% of individuals) were found in low scrub (including thickets and young second growth). In contrast, Yellow Warblers moving through Arizona (Hutto 1985a) and, especially, Kentucky (Mason 1979) were much less frequently found in that broad habitat type (80% and 39%, respectively). Hooded Warblers provide an even more striking example of geographic variation in use of stopover sites. In North Carolina and Kentucky, Hooded Warblers were never or rarely detected in old fields or thickets, being restricted primarily to tall forest habitats (Parnell 1969, Mason 1979). In contrast, along the Gulf coast of Mississippi and in Veracruz, Mexico, 80% of migrating Hooded Warblers were found in scrub habitats and avoided taller habitats (Moore et al. 1990, Winker 1995).

On the other hand, several species, such as Blue-headed Vireo (*Vireo solitarius*), Ovenbird (*Seiurus aurocapillus*), and Pine Warbler (*Dendroica pinus*), have not been shown to exhibit extensive geographic variability in habitat use during migration (compare Parnell 1969, Hutto 1985a, Mabey et al. 1993).

The lack of geographic consistency in habitat use by many migratory species suggests that migrants are adapted to exploit the unpredictable environments encountered along migratory routes (Morse 1971), and that the distribution of individuals across habitats is the result of complex, hierarchical evaluations of habitat suitability (Hutto 1985b, Moore et al. 1993; also see below). The wide variability in use of specific habitat types also highlights the limitations of using broad habitat categorizations for identifying priority habitats for individual species (Petit et al. 1993). For example, more detailed, quantified characterizations of habitats would allow better evaluation of vegetative features associated with particular species, which in turn could foster more consistent identification and effective management of stopover areas. Furthermore, if species are (at least partially) constrained in their use of habitat types during migration, for example by their morphology (Leisler and Winkler 1985; also see below), detailed characterization of habitat features will be necessary to understand the ecological and evolutionary basis of habitat selection.

Geographic variation in habitat use also could result from different ecological and physiological requirements that must be fulfilled along the migration routes. Stopover sites near breeding grounds, for example, may serve as refugia that allow individuals to complete prebasic molts; fat deposition may not be as critical (Cherry 1985, Winker et al. 1992a). In contrast, energetic considerations probably are of overriding importance for migrants using habitats adjacent to ecological barriers (Loria and Moore 1990, Bairlein 1991, Moore 1991a). Thus, the varied requirements of migrating birds may result in use of dissimilar habitats at different locations along migration routes.

ARE PATTERNS OF HABITAT USE CONSISTENT BETWEEN SPRING AND AUTUMN MIGRATORY PERIODS?

Seasonal differences in ecology, behavior, and physiology of migrating birds can be pronounced. For example, rates of movement during spring migration may be twice as high as those during autumn (Pearson 1990); many typically insectivorous species consume fruit during autumn, but not spring (Martin et al. 1951); continental migratory pathways can vary substantially between the two seasonal legs (e.g., "loop migration;" Cooke 1915, Berthold 1993); reproductive behavior is more pronounced during spring migration than during autumn (Quay 1985, Moore and McDonald 1993); and characteristics of fat accumulation may differ between the two periods (Blem 1980, Moreau 1969).

Seasonally related constraints or opportunities may influence, or be dictated by, patterns of habitat use. Hutto (1985a) observed significant seasonal shifts in habitat use by more than half of the 26 species that migrated during spring and fall through southeastern Arizona. Those shifts were highly correlated with changes in overall insect abundance. Blake (1984) documented substantial seasonal shifts by the avian assemblage migrating through Nevada, and concluded that changes may have reflected responses to a changing food base, or physiological constraints imposed by elevational factors. Likewise, Farley et al. (1994) studied migratory bird use of a successional gradient of riparian cottonwood stands in New Mexico. They found that, whereas species richness increased linearly with stand age during the spring, migratory birds preferred younger woodlands during autumn. In Iowa, several species of *Vermivora* that forage in trees during spring migration often are found in agricultural fields and weed patches during the autumn period (Dinsmore et al. 1984). Swainson's Thrush (*Catharus ustulatus*) and Northern Waterthrush (*Seiurus noveboracensis*) exhibited seasonally different patterns of habitat use while migrating through Minnesota (Winker et al. 1992a).

In contrast, data in Weisbrod et al. (1993) show that, when taken as a group, the migratory bird assemblage passing through the Saint Croix River Valley of Wisconsin exhibited similar proportional use of six habitats during spring and autumn. However, a pronounced increase in autumn use of the pine forest site was detected in that study (Weisbrod et al. 1993).

The above examples provide evidence of seasonal variation in habitat use by migrating birds, although only Hutto (1985a) and Yong et al. (1998) have systematically examined shifts at the species level. Indeed, seasonal changes in overall avian habitat use on a local scale may occur for several reasons unrelated to habitat shifts by species or individuals, such as high seasonal turnover of species (Lincoln 1935, Hutto 1985a), or seasonal changes in age structure of populations (Murray 1966, Ralph 1971). For example, Yong et al. (1998) found that patterns of habitat use by Wilson's Warblers (*Wilsonia pusilla*) migrating through New Mexico varied between spring and autumn and that those differences could be attributed to seasonal differences in the age and sex structure of the populations. Seasonal variation in habitat use also may be dictated by the ecological and physiological constraints unique to each season (see above). The extent and ecological basis of seasonal variation in use of migratory stopover habitats needs further study. In the meantime, seasonal variation in habitat use needs to be incorporated into conservation strategies.

ARE HABITATS USED DURING MIGRATION SIMILAR TO THOSE OCCUPIED DURING OTHER SEASONS?

Seasonally related patterns of avian habitat use (e.g., Rice et al. 1980, Collins and Briffa 1982) have profound consequences for wildlife management and conservation. Indeed, otherwise solid conservation efforts can be hampered because temporal changes in habitat use are not considered (e.g., Bancroft et al. 1992). To maximize effectiveness, management strategies for migratory populations should integrate not only summer and winter habitat requirements, but also those of migration periods (Moore and Simons 1992a, Petit et al. 1993). Delaying development of those plans, however, is a lack of information on the similarity of habitats used throughout different periods of the year.

Habitat use by neotropical migrants during the breeding season, and to a lesser extent the overwintering period, has been examined in detail relative to that during migration. Many species occupy superficially similar habitats in temperate breeding and neotropical wintering areas (Hutto 1985b, Petit 1991), although numerous exceptions also can be found (Rappole et al. 1983, Robbins et al. 1989b, Petit 1991). The similarity between migratory bird habitat use during migration and either the breeding or wintering season has not been thoroughly addressed. Because most conservation plans focus only upon breeding and wintering areas (Finch and Stangel 1993), such comparisons could serve to identify gaps in protection of important stopover habitats that are not encompassed by existing components of conservation plans.

Parnell (1969; also see Power 1971) observed that habitat relationships among 12 species of wood-warblers were consistent between migration and breeding periods in North Carolina. In that study, between-season overlap (for formula see Colwell and Futuyma 1971) in habitat distribution averaged 82% (SE = 2.5, range = 65–98%) for each species. Likewise, McCann et al. (1993) found that forest- and scrub-breeding species exhibited seasonal consistency in habitat use as they migrated through the coastal areas of the mid-Atlantic states.

In studies where the range of available habitats was more restricted, however, migrants used habitat types that were not characteristic of those used during breeding or wintering periods. For example, species migrating through coastal barrier islands of Mississippi occurred in habitats highly dissimilar to those used at other times of year, a phenomenon that Moore et al. (1990) attributed to lack of other, more preferred, habitats. Warblers that breed in deciduous forests exhibited strong habitat relationships while migrating through areas in Kentucky dominated by deciduous vegetation types (Mason 1979). In contrast, those species that nest in northern coniferous forests were more broadly distributed across vegetation types, suggesting less selectivity in those situations where preferred habitats are not present (Mason 1979). Most species passing through southeastern Arizona (Hutto 1985a) occupied an array of habitats at least superficially similar to those used during the breeding season.

The analysis conducted below (see WHAT ARE THE ECOLOGICAL CORRELATES OF HABITAT USE ALONG MIGRATION ROUTES?) demonstrates that species that occupy similar breeding habitats often are found together in the same habitats during migration. Furthermore, habitats used during those two periods are comparable in structural characteristics. In particular, species that breed in young successional growth tend to be found in scrubby areas and thickets during migration (Fig. 3). In Belize, Petit (1991) found that scrub-breeding migratory birds tended to overwinter in early successional habitats, whereas species that nested in taller forests were more generalized in their habitat distributions. In migration, forest-breeding species also tended to occur in the tallest habitats available, although as Petit (1991) suggested for overwintering birds, those species typically occur in a more diverse set of habitats than scrub-breeding species. Survey information from Parnell (1969), Moore et al. (1990), and Mabey et al. (1993) suggest that scrub-breeding species may be more restricted in habitat distributions during migration than are forest-breeding species (Fig. 4). In fact, species that nest in tall, forested habitats had an average niche breadth during migration that was 20% broader than those species that nest in younger successional habitats. That scrub-dwelling species make relatively limited use of the array of available habitats during migration indicates that some conservation efforts should focus on habitats of short stature because species that con-

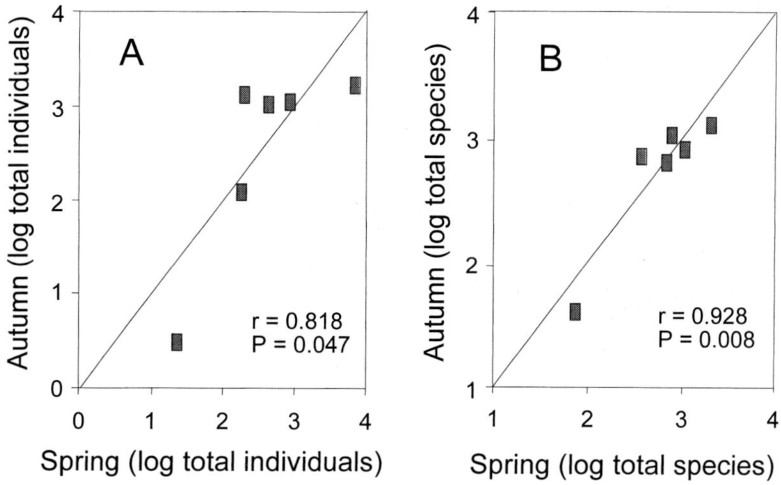

FIGURE 3. Comparison of seasonal use of six habitat types by birds migrating through Wisconsin summarized by relative numbers of (a) individuals and (b) species. Pearson correlation coefficients (r) were calculated from log-transformed raw values presented in Weisbrod et al. (1993; Figs. 2 and 3). Diagonal line represents identical habitat use between seasons.

centrate in them are less likely to use other habitat types. The apparent discrepancy between the preceding statement, promoting preservation of scrubby habitats for specialized species, and that made earlier advocating structurally diverse habitats to optimize species diversity, highlights the need for biologists to identify regional or other large scale conservation priorities before implementing local management plans for migrating birds (see L. J. Petit et al. 1995).

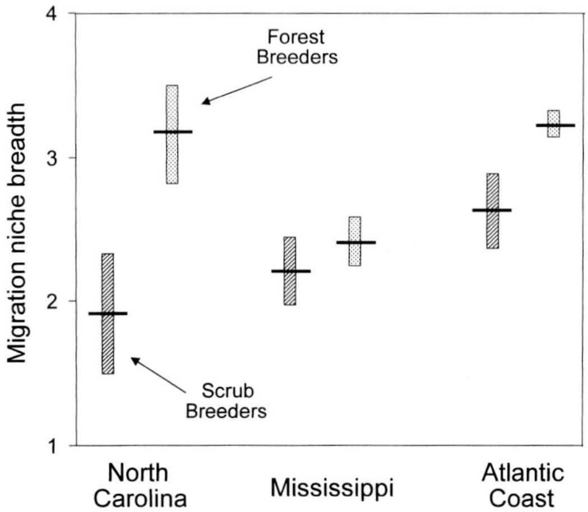

FIGURE 4. Comparison of habitat breadth (Levins 1968) during migration for species that nest in mature forest habitats (forest breeders) and species that nest in early successional habitats (scrub breeders). Horizontal line represents group average and vertical bar indicates one standard error. Data taken from Parnell (1969; North Carolina), Moore et al. (1990; Mississippi), and Mabey et al. (1993; Atlantic coast).

Several studies in Europe have demonstrated that, in general, species show greater variability in habitat use during migration than during either breeding or overwintering periods (Alatalo 1981, Bilcke 1984; but see Hansson 1983). In North America, Rice et al. (1980) also presented data in support of that pattern. Distributional data from regional works also show similar patterns. For example, approximately one-third of common nearctic-neotropical migrants that both migrate through and breed in California were identified by Zeiner et al. (1990) as occupying more habitat types during migration (Chi-square goodness-of-fit test; $\chi^2 = 28.7$, df = 2, $P < 0.001$); no species were more diverse during the breeding season. Likewise, of those nearctic-neotropical migrants noted to exhibit more diverse habitat use in one season or the other, 62% (8 of 13) in Missouri (Clawson 1982) and 64% (7 of 11) in North Carolina (Parnell 1969, Power 1971) were more diverse during migration than in the summer. Data from Stiles and Skutch (1989) indicate that, whereas 57% of the nearctic-neotropical migratory species that both migrate through and overwinter in Costa Rica did not exhibit noticeably different seasonal patterns of habitat use, 43% were more varied in habitat distribution during migration ($\chi^2 = 23.2$, df = 2, $P < 0.001$). Finally, observations by many amateur birdwatchers and bird banders during migration (e.g., Rudy 1971, DiGioia 1974) provide a wealth of anecdotal support for the above

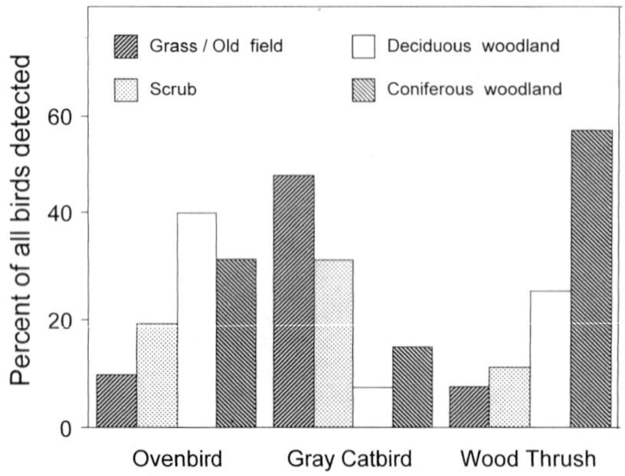

FIGURE 5. Year-round distribution of several species in different habitats in Pennsylvania (Davis and Savidge 1971). The data demonstrate the broad use of habitats by species that characteristically breed in mature forest (Ovenbird, Wood Thrush) and young successional growth (Gray Catbird).

generalization. For example, data from a year-round mist-netting effort in Pennsylvania (Davis and Savidge 1971) revealed that two forest-breeding species, Wood Thrush (*Hylocichla mustelina*) and Ovenbird, regularly occurred in brushy oldfields and other early successional growth, and approximately one-fifth of the scrub- and edge-nesting Gray Catbirds (*Dumetella carolinensis*) were captured in mature woodlands (Fig. 5).

Several patterns arise from the above observations. First, while seasonal variability in habitat use does occur, many species do have generally consistent, year-round affinities with broad habitat types. In particular, species that nest in relatively low vegetation types tend to use those habitats disproportionately during migration. Most exceptions to the generalization are detected in studies where species' preferred habitats are not available or are not incorporated into distributional surveys. The advantage of maintaining some consistency in habitat use throughout the year presumably relates to more efficient exploitation of those habitats for which the species is best adapted (Morse 1971, Greenberg 1984c).

Second, many long-distance, migratory species are capable of using a wide variety of habitat types during migration, some of which resemble neither their summer nor wintering habitats (Simons et al. *this volume*). Much of this apparent "indiscriminate" habitat use may be the outcome of a tradeoff between the cost of searching for higher quality habitats and the benefits of remaining in the already occupied habitat. From a theoretical standpoint, habitat assessment (and, hence, selection) should occur during migratory stopovers as individuals should be genetically predisposed or have the behavioral flexibility to locate those habitats that offer the greatest chances of survival (Hutto 1985b). Habitat switching by migrants has been observed at stopover sites (e.g., Moore et al. 1990) and early morning flights of nocturnal migrants heading inland along coastlines have been documented both in Europe and North America (Alerstam 1978, Wiedner et al. 1992). Other studies, however, have not detected significant reverse flights or movements between habitats (Bairlein 1983, Winker 1995). Regardless of whether local movements regularly occur or not, many individuals that settle into habitats in the early morning after nocturnal flights may have little option but to forage and rest in the locally available sites. For example, along coastlines where much of the native vegetation has been destroyed and where over-water flights were just completed, migratory birds may be "forced" into occupying the first habitats encountered because of depleted energy stores (Moore and Simons 1992a).

WHAT ARE THE ECOLOGICAL CORRELATES OF HABITAT USE ALONG MIGRATION ROUTES?

Data summarized in the preceding sections provide clear evidence of nonrandom use of habitat types by many species during migration. Although most species appear to be more generalized in habitat use during migration compared to other times of the year, the habitat-specific benefits and costs that are associated with the probability of completing the migratory journey must weigh heavily in the evolution of habitat discrimination. Animals should exhibit an affinity to those habitats that offer the greatest fitness advantages (Wecker 1964, Fretwell and Lucas 1970, Charnov 1976).

Several authors recently have addressed the issue of why nearctic-neotropical migrants might occupy certain habitats, and not others, during migration (Kuenzi et al. 1991, Moore and Simons 1992a). This section provides a brief overview and evaluation of several of those hypotheses. Understanding the ecological and evolutionary basis for habitat selection will ultimately lead to better management of migratory stopover sites and of long-distance migratory birds.

The distribution of birds among habitats during migration may be influenced by four features: (1) food abundance or effectiveness in exploiting the food base, (2) competition with other species, (3) predation pressure or relative safety from predators, and (4) reproductive opportunities

FOOD ABUNDANCE OR AVAILABILITY

If birds are choosing habitats during migration based upon food abundance or the ease with which food can be harvested, several (non-exhaustive) predictions can be made.

Bird abundance across habitats should be correlated with food abundance

Several studies have documented correlations between fruit or insect abundance and density of migrating birds through major habitat types (Hutto 1985a, Martin 1985) and through different areas within the same habitat (Blake and Hoppes 1986, Martin and Karr 1986). Terrill and Ohmart (1984) found that autumn migratory movements of Yellow-rumped Warblers in riparian woodlands of the southwestern United States were "facultative" and related to local abundance of insects.

These studies suggest that migrating birds respond to abundances of arthropods and fruit once settlement within a habitat has occurred or when the northern limits of wintering ranges are established during autumn. However, because only Hutto's (1985a) work involved surveys across more than two habitat types, the extent to which site-based food abundance influences avian habitat use during migration needs further examination.

Species that have relatively similar diets or that forage in similar ways should co-occur in the same habitats

Because foraging behavior of woodland birds during the breeding season is related to local vegetative structure (Robinson and Holmes 1982, Petit et al. 1990), migrating birds also might choose habitats with vegetative or other environmental features, including food abundance, that allow efficient gathering of food. If habitat use during migration is driven primarily by abundance of particular food resources or the ease with which those resources can be harvested, one would predict close concordance between habitat use and diet or foraging behavior, respectively. Data from five studies (Parnell 1969, Hutto 1985a [spring and autumn], Moore et al. 1990, Mabey et al. 1993) were used to test the hypotheses that dietary habits and foraging behavior are related to habitat types used during migration. Species were categorized by diet (omnivore or insectivore) and foraging location (canopy, shrub/understory, or ground gleaner, or aerial forager). Omnivores were defined as those species that include fruit or nectar as a major component of their diet (Ehrlich et al. 1988; pers. obs.).

To examine the above hypotheses, the relative use of the array of habitats was summarized for each species into a single index. For each study, principal component analysis (PCA; PROC PRINCOMP, SAS Institute 1990) was used to ordinate bird species by their proportional use of surveyed habitats. Scores on each principal component axis were derived for each species and were used to characterize habitat use by that species relative to all others. Thus, species with comparable patterns of habitat had similar scores along an axis. Only scores from the first two principal components were used in analyses as those two components accounted for more than half of the variation within all datasets ($\bar{x}$ = 67%, range = 53–85%). These scores were then used as dependent variables in a three-way analysis of variance (ANOVA) to evaluate differences ($\alpha = 0.10$) in habitat use between the two dietary guilds and among the four foraging guilds. Scores from the first PCA component were analyzed separately from the second component.

In addition to diet and foraging behavior, a third factor, breeding season habitat (each species categorized as breeding in either coniferous forest, deciduous forest, shrub, or edge/open habitats), was included in the ANOVA model. Although the relationship between habitat use during migration and the breeding season is evaluated separately below, foraging behavior was not independent of breeding season habitat (log-likelihood ratio [*G*] test; $P < 0.05$ in four of the five studies). Thus, inclusion of breeding season habitat use in the ANOVA models was appropriate to control for spurious relationships, and to provide a more comprehensive analysis of the correlates of habitat use during migration. Only main effects in the ANOVA model were relevant to testing of the above hypotheses (i.e., interactions were not examined).

Species that eat the same general type of food did not consistently co-occur in similar habitats during migration (Table 1, Fig. 6). However, data from the mid-Atlantic coast (Mabey et al. 1993, McCann et al. 1993) and Arizona highlands (Hutto 1985a) during autumn and along the Gulf coast (Moore et al. 1990) during spring provided some evidence that birds selected general habitat types based upon the types of food that were found there. In two of those studies (Moore et al. 1990, Mabey et al. 1993), omnivorous species tended to be overrepresented in scrub habitats and underrepresented in coniferous habitats. Insectivorous species exhibited varied responses to habitats across the five studies.

Other studies (e.g., Blake and Hoppes 1986, Martin and Karr 1986) identified food preferences as a strong correlate of habitat selection during autumn migration. One explanation for lack of a general relationship between diet and

TABLE 1. ECOLOGICAL CORRELATES OF HABITAT USE DURING MIGRATION

Study	Dependent variable	Source of variation	F	df	P
Parnell 1969	PC1	Overall	1.98	5, 13	0.15
		Diet	0.02	1, 13	0.88
		Forage	1.11	2, 13	0.36
		Nest	0.95	2, 13	0.41
	PC2	Overall	3.66	5, 13	0.03**
		Diet	0.01	1, 13	0.91
		Forage	1.83	2, 13	0.20
		Nest	7.77	2, 13	<0.01**
Hutto 1985a (Spring)	PC1	Overall	14.73	3, 18	<0.01**
		Diet	0.22	1, 18	0.64
		Forage	4.84	1, 18	0.04**
		Nest	7.11	1, 18	0.02**
	PC2	Overall	1.38	3, 18	0.28
		Diet	2.13	1, 18	0.16
		Forage	0.87	1, 18	0.36
		Nest	2.23	1, 18	0.15
Hutto 1985a (Autumn)	PC1	Overall	13.38	3, 22	<0.01**
		Diet	3.49	1, 22	0.08*
		Forage	0.71	1, 22	0.41
		Nest	13.74	1, 22	<0.01**
	PC2	Overall	0.35	3, 22	0.79
		Diet	0.71	1, 22	0.41
		Forage	0.28	1, 22	0.60
		Nest	0.28	1, 22	0.60
Moore et al. 1990	PC1	Overall	3.18	7, 12	0.04**
		Diet	0.20	1, 12	0.66
		Forage	2.60	3, 12	0.10*
		Nest	2.93	3, 12	0.08*
	PC2	Overall	6.38	7, 12	<0.01**
		Diet	3.53	1, 12	0.09*
		Forage	3.40	3, 12	0.05**
		Nest	1.23	3, 12	0.34
Mabey et al. 1993	PC1	Overall	3.43	7, 24	0.01**
		Diet	0.46	1, 24	0.50
		Forage	1.20	3, 24	0.33
		Nest	2.23	3, 24	0.11
	PC2	Overall	2.57	7, 24	0.04**
		Diet	6.03	1, 24	0.02**
		Forage	0.53	3, 24	0.67
		Nest	2.84	3, 24	0.06*

* $P \leq 0.10$; ** $P \leq 0.05$.

Note: For each of five studies, a three-factor analysis of variance (ANOVA) model was used to evaluate the effects of diet, foraging location (Forage), and habitat use during the breeding season (Nest), on habitat use during migration (PC1 and PC2). Only main effects are presented. See text for additional details.

habitat use in this analysis was that data from most of these studies were collected during spring migration when comparatively little fruit is available in North America. Mabey and coworkers (1993) collected their data in autumn, although Hutto (1985a) also worked during autumn in the Arizona desert and documented no obvious pattern of fruit-related habitat use (Fig. 6), at least at the scale at which habitat was measured. Stevens et al. (1977) found that the distribution of spring migrants moving through Arizona was closely related to insectivorous and granivorous food habits.

Foraging behavior was significantly related to habitat use in three of the five studies (Table 1, Fig. 7). Generally, species that glean insects from foliage in the shrub layer were underrepresented in pine-dominated forests, but were relatively abundant in shrubby habitats of low stature. Canopy foragers typically were most common in tall coniferous and broad-leaved forests. Thus, insect-gleaning species used those habitats that offered

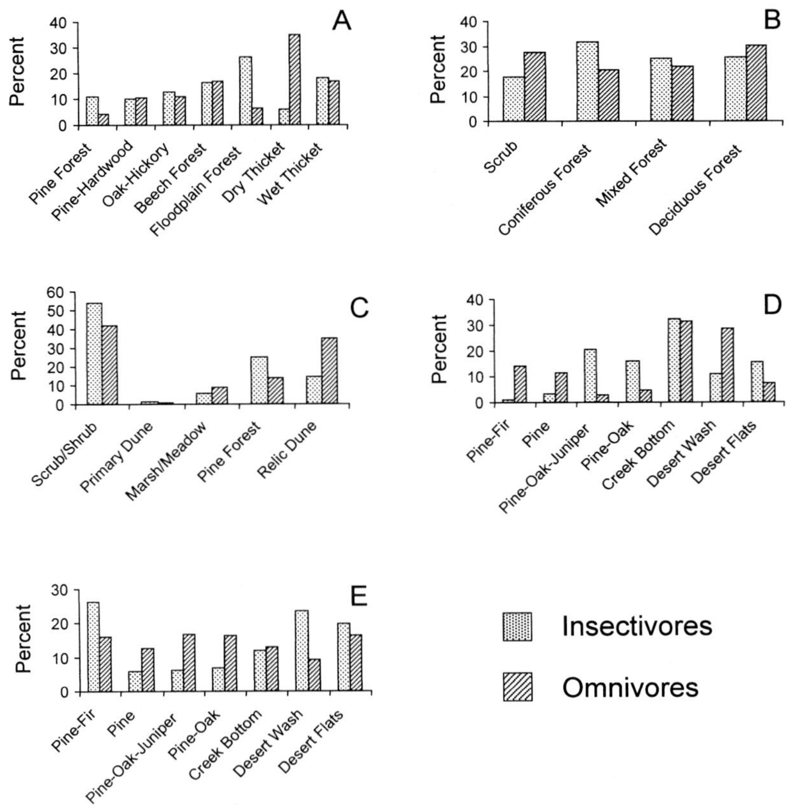

FIGURE 6. Distribution of omnivorous and insectivorous bird species among different habitat types during migration. Bars represent the average percentage of individuals for each species that was surveyed in each habitat type (thus all bars for each dietary category total 100%). Data are from (a) North Carolina (Parnell 1969), (b) mid-Atlantic coast (Mabey et al. 1993), (c) Mississippi coast (Moore et al. 1990), and southeastern Arizona (Hutto 1985a) in (d) spring and (e) autumn.

the densest foliage at preferred foraging heights. Ground foragers tended to use the tallest broadleaved vegetation available, whereas the few flycatching species represented in the data sets exhibited wide variability in habitat use.

Migratory species that breed in similar habitat types during the summer also occur together in a restricted set of habitats during migration (Table 1, Fig. 8). In all five studies, one of the principal components summarizing habitat use was significantly related to the patterns of habitat used during the breeding season. Thus, this analysis provides support for the assertion that long-distance migrants should occur in similar habitat types (if available) year-round because they are most effective in exploiting only a subset of environmental conditions (Morse 1971, Greenberg 1984c). It also supports the predictions of others that morphological constraints predispose species to select certain habitats over others. Several authors have demonstrated the relationship between morphology and use of habitat (including foraging behavior) during the breeding season (e.g., Miles and Ricklefs 1984, Leisler and Winkler 1985). Because morphology of a species is related to habitat use during the breeding season and remains relatively constant throughout the year, species that breed in similar habitats may also be found together at other times of the year, such as during migration. Bairlein (1992a) documented a close relationship between morphological characteristics and habitat distributions of neotropical migrants along the Gulf of Mexico coast during early spring, indicating that morphological constraints may influence habitat selection during migration (Berthold 1988).

Although habitat use during migration appears to be most closely related to vegetative characteristics of habitats used during the breeding season (Fig. 8), diet (Fig. 6) and foraging behavior (Fig. 7) may also influence habitats used during migration. The significant energetic demands of migration are believed to exert a

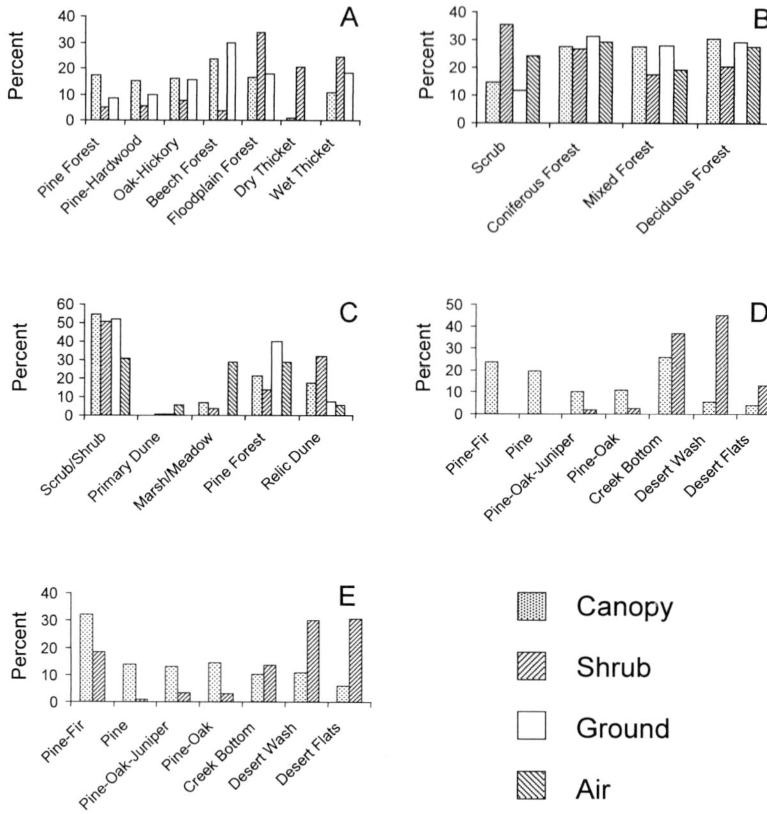

FIGURE 7. Distribution of canopy, shrub, ground, and air foragers among different habitat types during migration. Bars represent the average percentage of individuals for each species that was surveyed in each habitat type (thus all bars for each foraging category total 100%). Data are from (a) North Carolina (Parnell 1969), (b) mid-Atlantic coast (Mabey et al. 1993), (c) Mississippi coast (Moore et al. 1990), and southeastern Arizona (Hutto 1985a) in (d) spring and (e) autumn.

strong influence on the ecology, behavior, and evolution of migrating birds (Rappole and Warner 1976; Berthold 1975, 1993:92–106; Blem 1980, Moore 1991a). Thus, the lack of a close relationship between general dietary characteristics of species and habitat use during migration was unexpected. Many of the previous studies that have found a relationship between frugivorous behavior of migrating birds and habitat use have focused on microhabitat preferences within one habitat type (Blake and Hoppes 1986, Martin and Karr 1986), an analysis that was not possible using the data derived from the published reports used in this study. Other research has documented major patterns of habitat use during migration that were related to diet (Stevens et al. 1977, Martin 1985). Thus, although selection of habitats that offer the best foraging opportunities may partially account for the observed distributions of species during migration (Raitt and Pimm 1976, Martin 1980, Blake 1984, Hutto 1985a, Moore and Yong 1991), the analysis presented here suggests that affinities to broad habitat types used during the breeding season or other factors (e.g., predation pressure, morphological and energetic constraints, surrounding land-use patterns) may exert significant control over habitat selection (Hutto 1985b, Ward 1987, Moore et al. 1990). Distribution of individuals *within* those habitat types, however, may be more closely associated with abundance of insect and fruit resources (e.g., Blake and Hoppes 1986, Martin and Karr 1986). For example, Smith et al. (1998) suggested that Black-throated Green Warblers (*Dendroica virens*) foraging in forests within three kilometers of Lake Huron during spring migration used microhabitats (those closest to the water) that supported the greatest numbers of emerging aquatic insects.

COMPETITION

When settling into stopover habitats, migrating birds may respond to the combined effects of food abundance and the number of potential

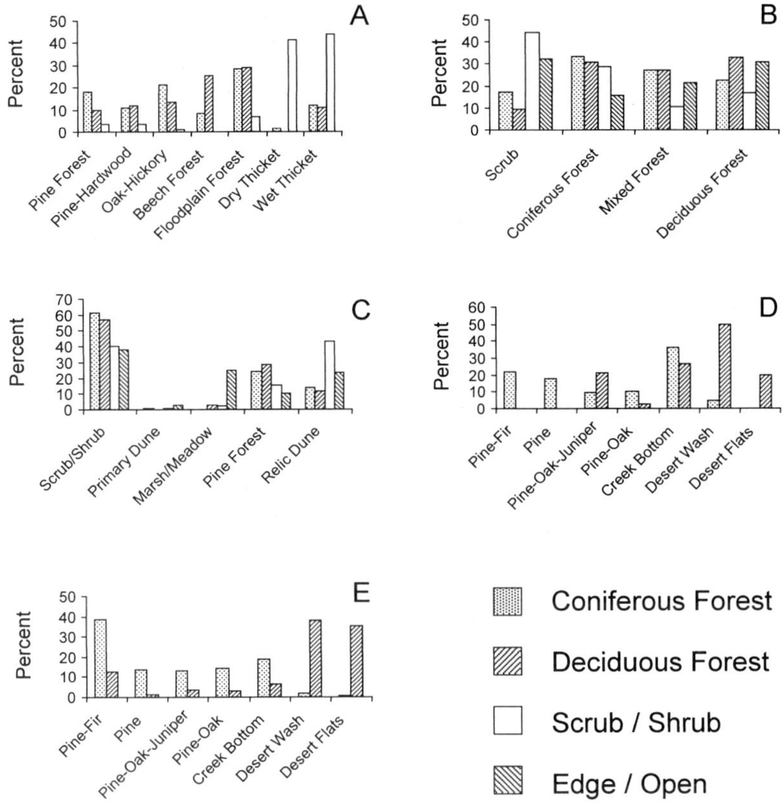

FIGURE 8. Distribution of coniferous forest-, deciduous forest-, scrub/shrub-, and edge/open-nesting species among different habitat types during migration. Bars represent the average percentage of individuals for each species that was surveyed in each habitat type (thus all bars for each nesting category total 100%). Data are from (a) North Carolina (Parnell 1969), (b) mid-Atlantic coast (Mabey et al. 1993), (c) Mississippi coast (Moore et al. 1990), and southeastern Arizona (Hutto 1985a) in (d) spring and (e) autumn.

competitors (both heterospecifics and conspecifics). Two facets of this hypothesis need to be verified. First, does competition among species or among individuals of the same species occur? And, second, if competition does exist, does it influence the distribution of individuals across broad habitat types?

Food-based competition occurs only when already limited resources are depleted by foraging individuals (Martin 1986). Abramsky and Safriel (1980) suggested that competition may have influenced the evolution of migration periods among Mediterranean species. Studies in Europe (e.g., Hansson and Petterson 1989; but see below) and North America (e.g., Martin 1980, 1981) have concluded that use of different habitat patches by transient species was determined (at least) partially through competitive interactions. Data on within-habitat resource use from several studies in Europe were consistent with expectations of interspecific competition (Ormerod 1990, Pambour 1990). Moore and Yong (1991) presented a brief synopsis of several studies that offered circumstantial evidence for interspecific and intraspecific (also see Greenberg 1986) competition among North American birds during migration. In perhaps the most convincing demonstration of *en route* competition, Moore and Yong (1991) found that, after having just crossed a 1,000-kilometer expanse of the Gulf of Mexico, the rate of mass gain by spring migrants on the Louisiana coast was influenced by the density of other small songbirds.

Other studies (e.g., Power 1971, Bairlein 1983, Fasola and Fraticelli 1990), however, have found that the distribution of migrating individuals was not consistent with the predictions of competition theory. In fact, Lovei (1989) concluded that competition for food resources among Palearctic-African migrants was minimal. Competition is most likely to occur at those locations where large numbers of migrants congregate near significant physical barriers, such as large bodies of water or small habitat patches surrounded by inhospitable habitat. Indeed, much of the evidence for competition among individuals has been generated at sites of relatively high avian density (e.g., Rappole and Warner 1976, Laursen 1978, Martin 1980, Moore and Yong 1991), while most of those studies not detecting competitive interactions have been conducted in areas where lack of physical barriers allow a more dispersed distribution of individuals (e.g., Power 1971, Fasola and Fraticelli 1990). Density-dependent intraspecific (Brown 1969, Fretwell and Lucas 1970) and interspecific (MacArthur 1972) interactions are believed to influence the distributions of birds among habitats. High relative densities of potentially competing species, such as on habitat "islands" or close to physical barriers, increase the probabilities that "interference" and "exploitation" competition (Schoener 1974) will occur.

These results have important ramifications for conservation of habitats along migratory routes. If migrating birds are most stressed after long flights over unsuitable habitats and in areas of high density, particular emphasis needs to be made for maintaining the ecological integrity of isolated stopover sites and sites near ecological barriers. Indeed, both ecologists (e.g., Rundle and Fredrickson 1981) and legislators (e.g., Public Law 99-645, the "Energy Wetlands Resources Act of 1986") have long recognized the importance of isolated stopover sites in the population dynamics of shorebirds, waterfowl, and rails. Conservation of large habitat patches in coastal areas, agricultural regions, and desert zones may serve to mitigate the detrimental effects of increased competition for resources in these areas.

PREDATION

Predation has been given very little attention as a factor shaping habitat use by migrating landbirds (Lovei 1989), largely because of the difficulties in documenting relatively rare predation events. However, in some situations bird-eating hawks and falcons, the principal predators of migrating birds, can cause substantial mortality (e.g., Lindström 1989, Moore et al. 1990, Alerstam 1993:343–344). Many birdwatchers also have observed a relationship between the timing of small landbird migration and that of their avian predators (e.g., Alerstam 1993:343).

The sparse data on predation in different habitat types makes difficult an evaluation of the hypothesis that habitat use or other behaviors of migrating landbirds is influenced significantly by predation pressure. Circumstantial evidence by Lindström (1990b) indicated that habitat use by migrating Bramblings (*Fringilla montifringilla*) in Sweden was a tradeoff between food abundance and predation pressure, a conclusion also supported for other species during the breeding and wintering seasons (Schneider 1984, Lima et al. 1987).

Several predictions can be made from the hypothesis that the behavior of migrants is influenced by predators along migration routes. If landbirds choose habitats or behave in ways that minimize the risk of predation, then one or more of the following relationships should be evident (data were extracted from Moore et al. 1990, the only published report that allows direct assessment of these hypotheses).

Either a negative or a positive relationship could exist between raptor and landbird abundance across local habitat types

Although Moore and Simons (1992a:351) stated that a positive relationship existed between predation pressure and migrant density, data from Moore et al. (1990) suggest that, in fact, the relationship is inconclusive ($r_s = -0.40$, $P = 0.60$, $N = 4$).

Migrants should exhibit disproportionate use of those habitats that afford the greatest protection from predators

Dense cover probably affords the greatest protection to small birds (e.g., Morse 1973, Grubb and Greenwald 1982; but see Lima et al. 1987). Thus, the shrub/scrub habitat in Moore et al.'s (1990) study probably offers the safest environment from predatory hawks, followed by pine forest, dune, and marsh/meadow. Data from Moore et al. (1990) support the above prediction ($r_s = 1.0$, $P < 0.05$, $N = 4$), although the value of these habitats as foraging sites may confound this relationship.

Species that make the greatest use of habitats that harbor high densities of raptors should exhibit more pronounced flocking behavior

By forming flocks, group members are believed to reduce their probability of predation (Pulliam 1973). Data from Moore et al. (1990; their Tables 2 and 4) provided little support for the prediction above. In the habitat with the highest relative density of bird-eating raptors, results opposite to the prediction were observed. That is, in pine forests, where raptor densities were greatest, a *negative* relationship ($r = -0.84$, $P < 0.01$, $N = 17$) was observed between percent overall use of that habitat and percentage of individuals of each of the 17 species that occurred in flocks. In contrast, migratory species that frequented scrub habitats (few raptors present), showed no relationship ($r = 0.17$, $P = 0.49$) between use of that habitat and propensity to join flocks.

Clearly, a simple relationship between predation pressure and habitat use does not exist. Rather, as other authors have noted, the primary consideration for birds during migration may be meeting energetic demands (Loria and Moore 1990, Moore 1991a). This does not imply that the risk of predation has not been important in the evolution of habitat selection by migrating birds, only that its potential importance is imbedded in a tradeoff between energetic gain and the risk of predation (or other factors). Furthermore, exact predictions are difficult to test using data that currently are available and, by lumping all migratory species together, important relationships between predators and prey may be obscured. Nevertheless, researchers need to continue to evaluate testable predictions in this area to fully understand habitat selection during migration.

Reproductive Opportunities

Several species of neotropical migrants are known to copulate while migrating in the spring, despite being up to 1,500 km from breeding areas (Quay 1985, 1989; Moore and McDonald 1993). Extra-pair copulation clearly could be beneficial to both sexes (Møller 1988, Westneat et al. 1990), although these benefits have not been empirically documented for birds that copulate during migration. Alternatively, *en route* copulation could occur between already paired birds (e.g., Greenberg and Gradwohl 1980), thus potentially minimizing the time required to lay a full clutch of eggs, and thereby maximizing the time to raise offspring on the breeding grounds. Female passerines can store sperm for more than 2 weeks and still produce viable eggs (Birkhead 1988).

The distribution of individuals among available habitats during the breeding season is believed to be based primarily on reproductive opportunities, while during migration replenishing energetic stores or protection from predators may be the primary selective force driving habitat selection (see above). However, given that some birds copulate during spring, habitat selection during migration also could be associated with potential reproductive benefits. If the "best" males are able to secure the highest quality stopover sites through an ideal dominance (i.e., despotic; Brown 1969) or other type of competitive interaction, females may reap reproductive benefits (in addition to food or predation benefits) by also occurring in those habitats. (A parallel argument also could be made, whereby males are attracted by the presence of females.) If territoriality or aggression (e.g., Rappole and Warner 1976, Bibby and Green 1980, Sealy 1988) among migrating males forces subordinate birds into lower quality habitats, and if *en route* copulation is beneficial to females, a basis for female (or male) preferences of certain stopover sites over others can be hypothesized. Although many males apparently are not physiologically capable of successful copulation during migration (e.g., Jones and Norment 1998), additional study is necessary to fully investigate the above ideas.

ARE GUIDELINES FOR MANAGEMENT OF SPECIES DURING THE BREEDING SEASON APPROPRIATE FOR MIGRATION PERIODS AS WELL?

Successful conservation of migratory species requires that temporal variation in habitat requirements be incorporated into management plans. The literature review and analyses above indicate that while many long-distance migratory species use superficially similar types of habitats during different stages of their annual cycles, substantial variation exists in this general theme. Specifically, many species appear to be more dispersed among available local habitats during migration than they are during the breeding season. This level of behavioral plasticity suggests that the same rigid guidelines (e.g., Robbins 1979, Faaborg et al. 1993) for conservation and management of breeding habitats may not be applicable to stopover habitats. These issues are reviewed below.

Recent bird conservation efforts in North America have focused on development of large-scale habitat management and conservation strategies (e.g., L.J. Petit et al. 1995). Thus, in addition to the historical approach of identifying important local habitat needs of species, the new strategies also incorporate landscape- and regional-level issues into local management directives. Wildlife managers now know that local populations cannot persist in isolation from the surrounding landscape (e.g., Rodiek and Bolen 1991). In this context, several issues are relevant to management of migration stopover habitats.

Fragment Size

Habitat fragmentation has detrimental effects on *breeding* bird populations (e.g., Lynch and Whigham 1984, Robbins et al. 1989a, Wilcove and Robinson 1990, Vickery et al. 1994), but a much less severe impact on overwintering nearctic-neotropical migrants (D. R. Petit et al. 1995). Unfortunately, little is known about the relative values of small and large habitat patches to *migrating* birds.

Yahner (1983) found no significant relationship between patch area and number of migratory species using small shelterbelts during spring in Minnesota. The sizes and range (0.2–

0.8 ha) of shelterbelts in Yahner's (1983) study, however, were small. In addition, few individual migratory species (e.g., Northern Oriole [*Icterus galbula*], Common Grackle [*Quiscalus quiscula*]) showed a preference for larger patches. In southern Wisconsin, Howe (1984) found that species richness and total density within small (<7 ha) survey plots in large forest tracts were similar to that recorded in nearby, small (<7 ha) forest fragments during spring and autumn migrations. Howe (1984), however, did not present information on species-specific responses to fragmentation.

Martin (1980) recorded a positive relationship between area and the number of species in shelterbelts that ranged from 0.1–3 ha in area. Density, however, was inversely related to fragment size. Individual species' distributions were not examined. Willson and Carothers (1979) found a strong positive correlation between island size and numbers of species migrating through isolated riparian forest patches along the Colorado River. In another study in southwestern United States, the number of species of springtime migrants was inversely related to area (and length) of woodland riparian corridors, although that relationship was not evident when total abundance of birds was examined (Skagen et al. 1998). Unfortunately, in this latter study several confounding factors such as elevation, isolation, and vegetative structure, may have obscured the true relationship between species use of woodlands of different size.

Several species migrating through northeastern Florida exhibited preferences for small (<5 ha) or large (>20 ha) maritime hammocks (Cox 1988). Long-distance migratory species that breed only in large forest tracts were detected disproportionately in larger patches, suggesting that species that are area-sensitive (sensu Robbins 1979) during the breeding season also prefer the largest available forest tracts during migration. For short- and long-distance migrants moving through the coastal plain of Maryland during spring, D.R. Petit et al. (unpubl. data) found mixed patterns for species that are considered area-sensitive or that typically nest in extensive forest tracts. For example, Ovenbirds and Yellow-rumped Warblers were more abundant in large (>300 ha) forest patches, whereas Black-throated Blue Warblers (*Dendroica caerulescens*) exhibited no such pattern (Fig. 9). Blackpoll Warblers (*D. striata*) were found most often in small (<150 ha) fragments. In Petit et al.'s study, habitat use during migration apparently did not reflect a simple relationship with patch size, but instead also appeared to be influenced by microhabitat characteristics and surrounding landscape (see below).

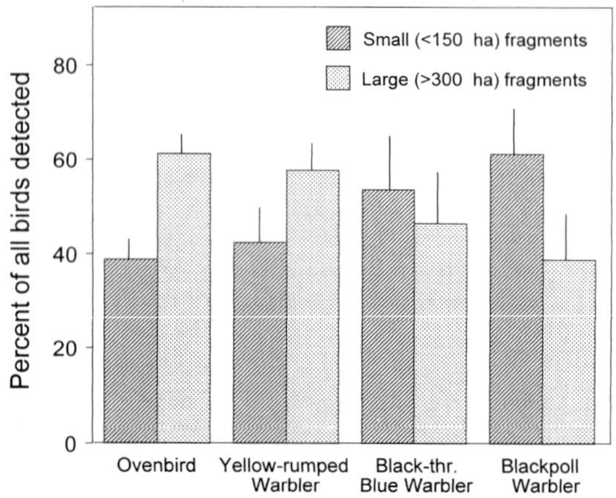

FIGURE 9. Distribution of Ovenbirds, Yellow-rumped Warblers, Black-throated Blue Warblers, and Blackpoll Warblers in 8 large (>300 ha; typically >500 ha) and 8 small (<150 ha; typically <80 ha) mature forest fragments in Maryland and Washington, DC, during spring migration. Bars represent the average percentage of all individuals for each species detected on a given day. Vertical line indicates one standard error.

Taken together, these studies suggest that maintenance of large tracts of relatively homogeneous, but structurally diverse, habitat is important for some species of landbirds during migration. However, constraints on habitat selection appear to be much more pronounced during the breeding season for most species. This suggests that habitat management guidelines developed for breeding birds will meet the requirements of most species during migration periods as well.

Isolation

By definition, habitat fragmentation results in isolation. The detrimental effects of isolation on animal populations has been both theoretically (MacArthur and Wilson 1967) and empirically evaluated (Shafer 1990, and references therein). However, for landbirds occupying terrestrial habitat islands (as opposed to oceanic islands), studies of the effects of isolation from similar types of habitat patches only recently have been investigated. Those studies conclude that some species are most likely to occupy forest fragments that are in close proximity to other, larger forest blocks (Lynch and Whigham 1984, Askins et al. 1987, Robbins et al. 1989a).

Little information exists to address this issue for migrating birds, however. Yahner (1983) detected an effect of isolation on birds migrating through agricultural shelterbelts of the upper Midwest. However, most species were more abundant in shelterbelts that were farther away from other woodlots. Martin (1980) concluded that degree of shelterbelt isolation was not related to species richness or abundance once habitat features and shelterbelt area were considered. Skagen et al. (1998) found that more species were detected in isolated oases compared to larger riparian woodlands in southeastern Arizona, although this pattern may have resulted from other, confounding factors (see above).

Corridors that connect fragments to larger habitat blocks have been suggested as remedies for decreased immigration and emigration often associated with wildlife breeding in isolated fragments (MacClintock et al. 1977, Wegner and Merriam 1979, Noss 1987). Several authors, however, have identified potentially serious ecological problems associated with presence of corridors (Whitcomb et al. 1981, Simberloff and Cox 1987).

During migration, corridors may provide a means by which individuals can readily find alternative stopover sites, as well as being "shunted" into appropriate breeding habitat during spring. D.R. Petit and co-workers (unpubl. data) surveyed more autumn transients in small forest fragments (<100 ha) that were connected via corridors than in similar-sized fragments that were isolated from surrounding woodlands. This same pattern was not observed during spring migration or for large (>300 ha) forest blocks in either season. These conclusions, while preliminary, suggest that wildlife corridors may enhance migratory bird use of small, isolated habitat fragments in some situations.

Given the wide diversity of results noted above with respect to the relationship between isolation and bird abundance, additional research is needed to evaluate the value of corridors to migrating (as well as breeding and overwintering) birds.

Local Habitat Diversity

Maximization of regional species diversity, as opposed to local diversity, is a guiding principle of conservation biology (Murphy 1989). For management of migratory birds, this typically means maintaining large habitat blocks necessary for sustaining viable populations of area-sensitive species and species susceptible to harmful edge effects (Faaborg et al. 1993). However, this rule may not need to be so strictly interpreted in management of habitats for landbirds during migration periods for three reasons: (1) migratory birds exhibit diverse patterns of habitat use during migration, so more species might be accommodated through local (landscape-level) habitat diversification; (2) many species appear to be capable of using a wide variety of habitats (compared to that used during the breeding season); and (3) the detrimental effects associated with small habitat fragments and edges (Wilcove and Robinson 1990) may not be applicable to transient individuals. Moore et al. (1993) also recommended that a diverse set of local habitats be maintained for birds during migration.

The contradiction between this recommendation and that suggested above for maintaining large habitat blocks for migrating birds is obvious. The solution, however, is to develop regional and local priorities for habitat types and species. For example, in regions where important breeding populations exist (e.g., Robinson et al. 1995), breeding season habitat requirements should be emphasized in local priorities. On the other hand, in locations where suburban or agricultural development has consumed all large blocks of habitat, conservation of existing small, interspersed patches might be promoted because of their benefits to migrating birds (Whitcomb et al. 1976). In addition, in regions where migrating birds may face particularly severe stress, such as areas bordering large physical barriers (e.g., Gulf of Mexico), local planners may desire to focus efforts on providing large blocks of high quality habitat, while not losing sight of the fact that a diverse set of stopover habitats would benefit a greater suite of species. Conservation benefits derived from local strategies and actions can be maximized only if developed within a larger, regional context.

CONSERVATION IMPLICATIONS

The discussions presented above clearly demonstrate the numerous sources of variation that are related to habitat use by migrating birds, including intraspecific, interspecific, geographic, behavioral, and seasonal elements. The complex nature of habitat selection, along with a paucity of well-designed studies to evaluate habitat use during migration, precludes formulation of firm management recommendations at the present time. However, by identifying some of the components influencing, or at least correlated with, habitat use, robust patterns are beginning to emerge. Documentation of these patterns is a critical step in development of detailed management plans in the future. The above analyses have several broad implications for management and conservation of migration stopover habitats.

(1) *The quality and importance of an area as a migration stopover site must recognize geographic location,* in addition to its vegetative, topographic, and other ecological characteristics. Specifically, habitat conservation and management for migrating birds must be given special attention in areas and along routes of heavy migratory bird movements (e.g., Atlantic coast),

and in areas adjacent to formidable ecological barriers (also see Moore et al. 1993, Petit et al. 1993, Cox 1995), such as large bodies of water (e.g., Gulf of Mexico, Great Lakes), and arid and highly agricultural regions.

(2) *Habitat patches surrounding ecological barriers must be of particularly high quality* because high densities of migrants at these sites may create a competitive environment with limiting resources. Habitat enhancement and restoration efforts need to focus on those concentration points to ensure sufficient food and shelter for prolonged occupancy by individual birds. Specifically, more extensive, undisturbed habitat should be maintained near barriers compared to areas where migrants are more spatially and temporally dispersed (Agard 1995). Where that may not be possible because of development, for instance, small landowner and backyard habitat programs may be effective in providing migratory birds critical habitats both before and after long flights over inhospitable barriers.

(3) *Habitat management and conservation priorities established in North America during the breeding season usually will be sufficient for providing the types, physical characteristics (e.g., patch sizes), and spatial arrangements (e.g., landscape connectivity) of habitats required by landbirds during migration periods.* Indeed, the behavioral plasticity exhibited by migratory birds during the spring and autumn, suggests that these species are able to effectively exploit widely divergent environmental conditions along migration routes. While any management generality will not apply to all species in all circumstances, these types of generalities offer land managers and planners a basis from which to begin to develop management plans. Needless to say, all conservation generalities need to be adapted to local situations.

(4) *Maintenance of relatively tall and structurally diverse forest types should be a high priority in stopover habitat management plans* because structurally diverse habitats generally support greater numbers of migratory species than habitats of low stature or vegetative complexity. Again, while this type of recommendation may be appropriate for most species, certain species may require different management actions.

(5) *Landscape-level and regional conservation plans should ensure a diversity of habitats for migrating landbirds.* The above recommendation notwithstanding, managers must also consider species with habitat requirements that do not include taller forests. During migration, early successional and grassland species appear to be more confined to habitat types that mimic those used during the breeding season, than are species that nest in mature forests. Thus, grassland and scrub habitats should be closely managed and positioned in ways that do not diminish their own quality or that of adjacent forests; for example, creation of a "checkerboard" of small habitat patches is not recommended.

(6) *Although the quality of small habitat fragments probably is inferior to large patches, small parcels should be protected as "migration stepping stones" (Date et al. 1991) when possible, especially in the absence of large habitat patches (e.g., shelterbelts, suburban parks).* In addition to the ecological benefits, parks offer an opportunity for people to view (and hear) the grand phenomenon of migration.

(7) *Local planning for management of migratory bird stopover habitats must consider and integrate both landscape- and regional-level issues.* Decisions for site-level conservation actions should not be made without consideration of landscape-level processes (e.g., plant and animal population dynamics) or patterns (e.g., composition and spatial distribution of landscape elements). Furthermore, optimal site-level conservation strategies can be achieved only by viewing objectives and biological targets in a regional context.

FUTURE RESEARCH

Good management decisions cannot be made in the absence of sound biological information. The lack of attention by scientists to migratory landbird habitat use during spring and autumn has severely hindered the ability of land managers to preserve the ecological integrity of migration stopover habitats. For research to contribute fully to management and conservation of migratory birds, a comprehensive strategy must be devised to understand the complexities of migration, including the underlying evolutionary, behavioral, and ecological components associated with the migration phenomenon. Moore (1991b) identified three broad arenas that require more in-depth study: (1) the evolutionary correlates of bird migration; (2) the energetic and ecological costs associated with travelling long distances over hostile environments and through unfamiliar habitats; and (3) the factors regulating the population dynamics of migratory birds, not only during the breeding and overwintering seasons, but also during migration. In addition, more thorough investigation of the seven main questions addressed in this review paper is urgently needed. Habitat requirements of individual species, as well as season-, age-, or sex-related variability in those patterns, need to be evaluated through rigorous surveys and habitat association studies. And finally, for conservation planners to establish a regional framework for preservation of critical migration stopover habitats, research ecologists must establish criteria by which important sites can be identified and managed. Without rigorous, biologically based standards, important stopover sites will "slip through the cracks," and with those habitats will go migratory birds.

ACKNOWLEDGMENTS

K. E. Petit and L. J. Petit have helped shape my thoughts about habitat selection of migratory birds. I thank D. Ewert, J. Faaborg, W. C. Hunter, R. L. Hutto, F. R. Moore, L. J. Petit, J. T. Rotenberry, and J. L. Trapp for taking the time to suggest numerous ways to bolster the strength and presentation of the ideas in this manuscript. D. Ewert, B. D. Watts, and S. E. Mabey graciously provided preliminary reports of their work. The late J. F. Lynch engaged me in numerous discussions on migratory bird ecology, and the Smithsonian Environmental Research Center provided a professional environment in which to begin development of my migration work.

Studies in Avian Biology No. 20:34–42, 2000.

MECHANISMS OF *EN ROUTE* HABITAT SELECTION: HOW DO MIGRANTS MAKE HABITAT DECISIONS DURING STOPOVER?

FRANK R. MOORE AND DAVID A. ABORN

Abstract. Evidence reveals that habitat selection occurs during migration, but little is known about how migrants made decisions about habitat use during stopover. Although most nocturnally migrating birds end their migratory flight well before dawn, selection of habitat probably occurs during daylight hours. Possibly "morning flights" represent efforts to explore suitable habitats in which to rest and forage. Choice of habitat probably consists of a sequence of hierarchically ordered decisions that depend on different criteria. When migrants arrive they may settle in response to gross habitat features such as vegetation density, then search for resources based on more subtle habitat features or the behavior of other migrants. We used radio-telemetry to study the movement pattern of Summer Tanagers (*Piranga rubra*) during stopover. Use of habitat and pattern of movement differed between fat and lean individuals. Lean birds were more active, displayed a pattern of movement more consistent with exploration, and visited more habitat types than did fatter birds.

Key Words: exploration, habitat selection, landbird migrants, migration, stopover.

"The crucial first step to survival in all organisms is habitat selection. If you get to the right place, everything else is likely to be easier."—E. O. Wilson. *Biophilia.* 1984

Habitat is an area possessing resources and other environmental attributes that promote occupancy and survival of individuals (Morrison et al. 1992). Habitat selection may be defined (Block and Brennan 1993) as "innate and learned behavioral responses of birds that allow them to distinguish among various components of the environment resulting in disproportional use of environmental conditions to influence survival and ultimate fitness of individuals." This definition identifies habitat selection as a process with fitness consequences (Hildén 1965, Klopfer and Hailman 1965; see also Hutto 1985b).

When a migratory bird stops *en route,* it almost invariably finds itself in unfamiliar surroundings when energy demands are likely to be high (e.g., Loria and Moore 1990, Martin and Karr 1990, Moore 1991a). It may also face conflicting demands between predator avoidance and food acquisition (e.g., Metcalfe and Furness 1984, Lindström 1990b, Moore 1994) and competition with other migrants and resident birds for limited resources (e.g., Hutto 1985a, Moore and Yong 1991). How well migratory birds satisfy energetic requirements and meet contingencies that arise during passage depends largely on their ability to locate resources and avoid sources of stress (sensu Jander 1975).

If selection of one habitat over another during stopover has consequences for a migrant's survival and subsequent reproduction, then migrants should display a preference for certain habitats and select among alternatives during stopover (Hutto 1985b; Moore et al. 1990, 1995). Three lines of evidence reveal that habitat selection occurs during migration (Petit *this volume*): (1) year-to-year constancy in species-specific patterns of distribution among different habitats (Bairlein 1983); (2) seasonal shifts in habitat types (Winker et al. 1992a, Weisbrod et al. 1993, Moore and Woodrey 1993), sometimes correlated with changes in food availability (Hutto 1985a, Martin 1985, Martin and Karr 1986); and (3) use of habitat out of proportion to its availability (Johnson 1980; e.g., Moore et al. 1990, Bruderer and Jenni 1990, Mabey et al. 1993). Migrants that assess habitat quality should gain an advantage relative to individuals that fail to assess habitat options *vis-a-vis* choice of habitat (see Hutto 1985b, Moore et al. 1995).

Study of habitat selection in migratory birds has focused largely on describing habitat use during stopover (e.g., Bairlein 1983, Moore et al. 1990, Winker et al. 1992a) rather than examination of the underlying proximate mechanisms that mediate selection. How do migrants distinguish one habitat from another? How is habitat quality assessed? What cues do migrants use when deciding to settle in a particular habitat? These are proximate questions about the mechanisms of habitat selection, rather than questions about the functional consequences of habitat choice (Hutto 1985b).

How birds assess habitat is less clear than their decision-making about individual resources, and the mechanisms used to make habitat choices are best known for decisions made outside the migratory season (Hildén 1965, Klopfer and Hailman 1965, Morse 1985, Morton 1990). We are only beginning to understand migrant-habitat relations during migration (see Hutto 1985b; Moore et al. 1990, 1995), much less appreciate the mechanisms migrants use to identify habitat attributes on which habitat choices are made during passage. Our objective is to explore how migrants might select habitat during stopover and to suggest ways to test predictions regarding habitat assessment.

To understand how migrants assess habitat during passage, it must be realized that birds make decisions at different spatial scales and that different factors, some extrinsic to habitat per se, operate at these different scales (see Hutto 1985b, Moore et al. 1995). Intrinsic constraints on habitat use are those factors thought to determine habitat quality and upon which migrants made decisions about habitat use. As the spatial scale broadens, factors intrinsic to habitat give way to factors largely unrelated to habitat, such as synoptic weather patterns. Yet, extrinsic factors may constrain opportunities to select habitat, not to mention the process of assessment itself.

Migration in relation to the Gulf of Mexico illustrates how an extrinsic factor such as prevailing winds constrains habitat use at different spatial scales. The likelihood of a successful flight across the Gulf of Mexico is tied to the occurrence of favorable flight conditions (Buskirk 1980). In spring, the peak of trans-Gulf migration, which occurs over the latter half of April through early May, corresponds to a period of predictable southerly airflow. Should migrants encounter unfavorable weather, flight is prolonged and energy stores depleted. Habitat assessment is likely to vary with energetic condition upon arrival (see Moore and Simons 1992a). Although migrants are observed crossing the Gulf of Mexico in fall, prevailing weather conditions during the peak of fall migration along the northern coast of the Gulf of Mexico facilitate movements parallel to the coast rather than across this barrier (Able 1972, Buskirk 1980; see Sandberg and Moore 1996). As a consequence, migrants are likely to encounter habitats during fall passage that differ from the habitats experienced during the return passage in spring.

WHEN DO MIGRANTS SELECT HABITAT?

Most nocturnally migrating songbirds end their migratory flight well before dawn (Kerlinger and Moore 1989), although exceptions occur, especially when night migrants must cross water barriers (see Gauthreaux 1971, 1972; Moore and Kerlinger 1991) or deserts (Moreau 1972; but see Biebach 1985, Bairlein 1987b). When crossing bodies of water, migrants have little choice but to continue migration until "suitable" habitat is found.

When migrants end their flight at night, selection of habitat probably occurs during daylight hours, most likely early in the morning, and not at night when landfall occurs. Although the behavior of nocturnally migrating birds is influenced by gross topographical features such as rivers (Bingman et al. 1982), mountains and valleys (Bruderer and Jenni 1990), and coastlines (Able 1972), migrants probably do not possess the sensory capability to evaluate subtle habitat differences at night (cf. Martin 1990). If visual capacity constrains decisions about habitat upon nighttime landfall, we would expect the distribution of migrants among habitats and the amount of movement to vary in relation to night-time light levels (e.g., moonlit versus overcast night). We would also expect to observe age related differences in the interpretation of habitat when making landfall at night (see Gauthreaux 1982a). Little is known about habitat decisions immediately upon landfall, even when arrival takes place during daylight hours (but see Gauthreaux 1972, Moore et al. 1990).

Nocturnal migrants have been observed making "morning flights" at several locations in North America (Bingman 1980, Hall and Bell 1981, Wiedner et al. 1992; D. Cimprich, unpubl. data) and Europe (Alerstam 1978, Lindström and Alerstam 1986, Spina and Bezzi 1990). These morning flights appear to differ from normal nocturnal migration in that (1) they occur during daylight usually within the first two hours after dawn, (2) they occur at low altitudes (sometimes from treetop to treetop), (3) flights are of short duration, and (4) migrants are often in flocks. Moreover, the direction of "morning flight" is not necessarily the same as the previous night's migration (e.g., Wiedner et al. 1992; but see Bingman 1980), although there is no a priori reason why migrants would not bias their daytime movement in the migratory direction. For example, trapping data for fall migrating Sedge Warblers (*Acrocephalus schoenobaenus*) in northern Italy reveal movement concentrated in early morning (Spina and Bezzi 1990), which suggested that newly arrived birds were moving away from the landing area. When juvenile and adult Sedge Warblers were tested in orientation cages at dawn, juveniles showed more intense but less directional activity, whereas adults were oriented along the migratory direction (Spina and Bezzi 1990). Other observations suggest that birds may engage in "morning flight" to compensate for drift experienced during nocturnal migration (Moore 1990a).

These morning flights may represent a period of exploration as migrants seek more suitable habitat in which to rest and forage (see Lindström and Alerstam 1986, Wiedner et al. 1992). If so, the distribution of migrants among habitats should change between the time of arrival and subsequent settlement (i.e., migrants should distribute themselves according to variation in habitat quality during morning flight). Although this expectation has not been tested, some observations are consistent with an "exploratory" function for these morning flights. On the Cape May peninsula, New Jersey, thousands of fall migrants can be observed in "morning flight" to the north, away from the end of the peninsula, toward the forested areas up the Delaware Bayshore (Wiedner et al. 1992). At other sites in the New Jersey coastal plain, "morning flight" is to the west or northwest, again toward forested areas (S. Gauthreaux, pers. comm.). Once the birds reach forested areas they diffuse, presumably settling in preferred habitats. When migrants stopover in alpine areas while crossing the Alps, juveniles are more likely to land in unfavorable habitats than adults (Bruderer and Jenni 1988). Finally, landbirds seem to be "attracted" to riparian areas following a night's migration in the southwestern United States (Terrill and Ohmart 1984). Within a riparian system consisting of different habitat types that presumably vary in suitability (Finch and Yong *this volume*), migrant abundance could be estimated relative to habitat availability and body condition assessed through mist-netting activity (J. Kelly et al., unpubl. data).

INFORMATION USED TO ASSESS HABITAT

Choice of habitat during passage probably consists of a sequence of hierarchically ordered choices (Johnson 1980; see also Orians and Wittenberger 1991), and different criteria may assume importance at different stages. Upon arrival, a migrant might settle on basis of gross habitat features (e.g., vegetation density), possibly choosing (or avoiding) area that bears some resemblance to previously experienced habitat. The decision to actually search for resources within a circumscribed area or to continue local movement may depend on more specific habitat features or the behavior of other migrants. During exploration a migrant might sample resources to make a more refined assessment of habitat.

Actually, the information migratory birds use to select among alternative habitats during stopover and the manner in which they do so is poorly understood. Migrating birds use habitat *en route* in different ways for different reasons: some birds try to deposit fat stores, others use the site as a molting ground (e.g., Winker et al. 1992a), while other birds simply rest until nightfall (e.g., Biebach 1990). Moreover, a hungry, fat-depleted individual undoubtedly has a different perspective on habitat than a migrant that is simply looking for a safe place to rest, and consequently may react to different habitat features when selecting habitat.

Because food is critically important *vis-a-vis* energetic requirements, we consider habitat cues from the perspective of a migrant searching for food resources. What are the environmental stimuli that evoke a settling response in relation to habitat during stopover? Although the proximate factors that serve as "cues" in habitat selection are probably linked to the ecological factors that determine suitability (see Hutto 1985b, Moore et al. 1995), they are not necessarily those that directly affect the migrant's survival in a given habitat (Lack 1933, Klopfer and Hailman 1965).

VEGETATION STRUCTURE

The "niche-gestalt" model of habitat selection (James 1971, James et al. 1984) would predict that migratory birds respond to gross structural features (e.g., vegetation density or height), presumably because those features are correlated with feeding demands. Structural features of breeding habitat (Sherry and Holmes 1985, Morse 1989), including microhabitat structural components such as needle architecture (Parrish 1995a), do influence dispersion of birds in forest habitats. Results of a foliage discrimination experiment (Greenberg 1985) showed that migrant wood warblers (*Dendroica castanea* and *D. pensylvania*) rapidly learn to distinguish between leaf-types, which may explain why foliage-gleaning birds develop preferences for particular plant species.

If foraging success varies with vegetation structure (e.g., Robinson and Holmes 1984), energetically constrained migrants should attend to structural features as cues when selecting habitat during stopover. When Hutto (1985a) examined the distribution and abundance of migratory species over an elevational habitat gradient in the Chiricahua Mountains, the pattern of habitat used differed between spring and fall, yet few vegetation variables changed seasonally within habitats. Hutto (1985a) concluded that these habitat variables are unlikely to be the proximate cues used by the birds for a settling response. This expectation could be tested more directly by presenting migrants with alternative "habitats" that varied in some structural feature thought to be important in habitat assessment. Although it would be possible to arrange such an experiment in the field (e.g., Gwinner et al. 1985), the more manageable, albeit contrived, approach would be examine habitat preference in an aviary setting where greater control can be achieved (e.g., Hebrard 1978, Partridge 1979, Roberts and Weigl 1984, Morton 1990). We would not be surprised if migrants responded to simple structural features, such as verticality (e.g., Morton 1990), which would provide flexibility when selecting habitat during migration.

SOCIAL ATTRACTION

Migrants may respond to the presence of other migrants, especially conspecifics, rather than habitat per se, when assessing habitat during stopover. Presumably a more suitable habitat would attract more individuals (sensu Fretwell and Lucas 1970; see Moore and Simons 1992a), although more migrants would more rapidly deplete resources and increase the likelihood of competition (Lindström et al. 1990, Moore and Yong 1991). In any case, responding to the behavior of other migrants as a mechanism to assess habitat quality would be especially useful if an area could not be thoroughly searched because of time constraints.

Imitation and observational learning are known to influence food preferences (e.g., Murton 1971, Tramer and Kemp 1979) and the effect is amplified when birds are food deprived (Mason and Reidinger 1981). When migrants stopover they often find themselves in the presence of other migrants, many of whom are trying to re-gain depleted nutritional stores. Circumstantial evidence leads us to believe that social facilitation may be involved in habitat selection during stopover. During "fallouts," when large numbers of migrants land at a stopover location over a short period of time, we have observed migrants to move quickly from tree-top to tree-top or shrub to shrub among habitats, often in loose mixed-species flocks, giving the impression that they were assessing habitats. Once migrants "settled" in a habitat, which appeared to take place within one or two hours after fallout, they often foraged alone or in small homospecific flocks (Moore et al. 1990).

If migrants "cue" on one another when selecting habitat, we would expect migrants to respond to playback of conspecific vocalizations (D. Cimprich, unpubl. data) or to the presence of model birds (cf. Gotmark and Unger 1994). Whether a migrant is attracted or repulsed by the presence of other migrants may depend on the bird's foraging strategy and the economics of resource defense during stopover. For example, Gray Catbirds (*Dumetella carolinensis*) rely heavily on patchily distributed fruits during fall passage (C. Dwyer, pers. comm.) and may attempt to exclude other migrants from those resources. Moreover, migrants that "cue" on the number of other migrants present in a habitat may also attend to the level of activity (e.g., feeding behavior) of other individuals as an indication of habitat use.

PREDATION RISK

Predation constitutes a significant hazard to migrating birds (e.g., Rudebeck 1950, 1951; Walter 1979, Kerlinger 1989, Lindström 1989, Moore et al. 1990). Consequently, the decision to use a particular habitat is influenced by the migrant's perception of predation risk (Lindström 1990b). Moreover, the need to avoid predators must be balanced against the need to acquire food to meet the energetic demands of migration. Balancing conflicting demands is not easily achieved during stopover for several reasons (Cimprich and Moore, in press): risk of predation varies widely in time and space (Aborn 1994); migrants are probably unfamiliar with foraging opportunities and predation risk; and they are under pressure to travel quickly. The reaction to predation and resolution of a conflict between food acquisition and predator avoidance surely varies with habitat (e.g., Lindström 1990b) and behavior of the predator (cf. Curio 1993), as well as the migrant's condition (Moore 1994) and age (Metcalfe and Furness 1984; D. Cimprich, unpubl. data).

Given time and energy constraints, we expect migrants to be especially sensitive to the threat of predation during stopover; time devoted to anti-predator behavior when there is little threat of predation is time not spent satisfying energy demand. Because the threat of predation is highly context dependent during stopover, we expect migrants to use various mechanisms of risk assessment; that is, a migrant behaves as if it were monitoring the degree of threat to itself or to other migrants (see Curio 1993).

SAMPLING RESOURCES DIRECTLY

A migrant might respond to resources independent of habitat structure or social factors. Information could be gathered about habitat by sampling resources directly, which might include the number of food items consumed, the time spent in a habitat, or the time since the last food item was consumed. A direct, accurate assessment of resources or sources of stress may be especially important to migrants because of the unpredictable nature of passage.

Because migrants find themselves in unfamiliar surroundings during stopover, particularly hatching-year migrants on their first passage, how they respond to novel circumstance will affect their use of habitat. Habitat decisions predicated on sampling could be constrained by a migrant's readiness to approach and feed on new food sources or in novel situations (sensu Greenberg 1984b, 1990). Although neophobia varies among species (Greenberg 1984b), the degree of neophobia does not decrease with increased hun-

TABLE 1. AVERAGE (±1 SD) MEASURES OF MOVEMENT FOR FAT AND LEAN SUMMER TANAGERS ON HORN ISLAND, MISSISSIPPI, 1992–1994

Movement variable	Fat (N = 11)	Lean (N = 8)	t-value	P
Linear distance (m)	229.2 ± 130.6	491.5 ± 380.4	2.11	< 0.05
Total distance (m)	1,003.8 ± 740.2	1,513.9 ± 831.3	1.16	> 0.05
Time between moves (min)	28.5 ± 25.1	18.0 ± 16.9	2.75	< 0.05
Length of move (m)	68.9 ± 24.8	89.2 ± 63.4	3.03	< 0.05
Rate of movement (m/min)	6.5 ± 3.1	12.1 ± 9.6	2.87	< 0.05

ger (Greenberg 1987b), which suggest that responsiveness to novel circumstances may not change with heightened energy demand during migration (but see Loria and Moore 1990). Likewise, social enhancement does not reduce feeding neophobia in Chestnut-sided Warblers (*Dendroica pensylvanica*; Greenberg 1987b; but see Coleman and Mellgren 1994).

PRIOR INFORMATION

The disadvantage of sampling habitats is that it takes time, a commodity presumably in short supply for most migrants (Alerstam and Lindström 1990). Pressure to arrive at a destination in a timely manner probably explains why most migrants depart the night of their arrival day, if not sooner (Winker et al. 1992b,c), although length of stopover does vary with the migrant's energetic condition (Moore and Kerlinger 1987, Kuenzi et al. 1991, Morris et al. 1994). Sampling may also expose migrants to increased risk of predation. Nevertheless, previous experience in different locations is known to influence later choices (Partridge 1979), so migrants might arrive at a stopover site with prior information of the distribution of resources (or sources of stress) in the environment, which would increase efficiency with which they use habitat (cf. Valone 1992).

Prior information would include any information about habitats gathered prior to arrival at a given stopover site, including experience with habitats on the breeding grounds, wintering areas, and previous stopover locations. Black-chinned Hummingbirds (*Archilochus alexandri*) rely on prior information when making foraging decisions during stopover, although the extent to which they do so depends on environmental variability (Valone 1992). High fluctuations in resource abundance and availability makes the use of prior information impractical.

The difficulty with using prior information gathered during migration is confounded by the fact that migrant landbirds seldom stop over at the same location. Suppose a migrant's settling response is influenced by previous experience, and the nature of that experience influenced by extrinsic factors (e.g., synoptic weather). As a consequence of extrinsic constraints, a migrant finds itself searching for food in a "strange" habitat, yet that experience influences subsequent decisions about habitat use.

INNATE PREFERENCES

Some habitat decisions made during passage are probably affected by innate (programmed) preferences (Hildén 1965, Klopfer and Hailman 1965), yet little, if any, attention has focused on the importance of such preferences during migration. Although it is unlikely that migrant landbirds would possess programmed habitat preferences specific to the migratory period, they may reference innate information about breeding habitat when making habitat decisions during passage. It is not unusual for migrants to occupy habitat during the nonbreeding season that resembles their breeding habitat (e.g., Parnell 1969, Power 1971, Lack and Lack 1973). Such behavior, which is consistent with the existence of innate preferences, may be especially beneficial for hatching-year birds given their lack of experience with different habitat types. Bairlein (1983) found species-specific habitat use among passage migrants that stopover at the west end of Lake Constance, Germany, and interpreted the year-to-year consistency in habitat distribution among adult and young birds without knowledge of the stopover area as evidence of innate preferences. If migrants do display innate habitat preferences during passage, we would expect *en route* experiences to shape those preferences in the face of variability encountered during passage.

→

FIGURE 1. Radio locations and movement tracks of lean (a, b) and fat (c, d) Summer Tanagers radio-tracked on Horn Island, Mississippi, following migration across the Gulf of Mexico. Two letter acronyms refer to habitat types: MM (marsh/meadow), PF (pine forest), RD (relic dune), SS (scrub/shrub).

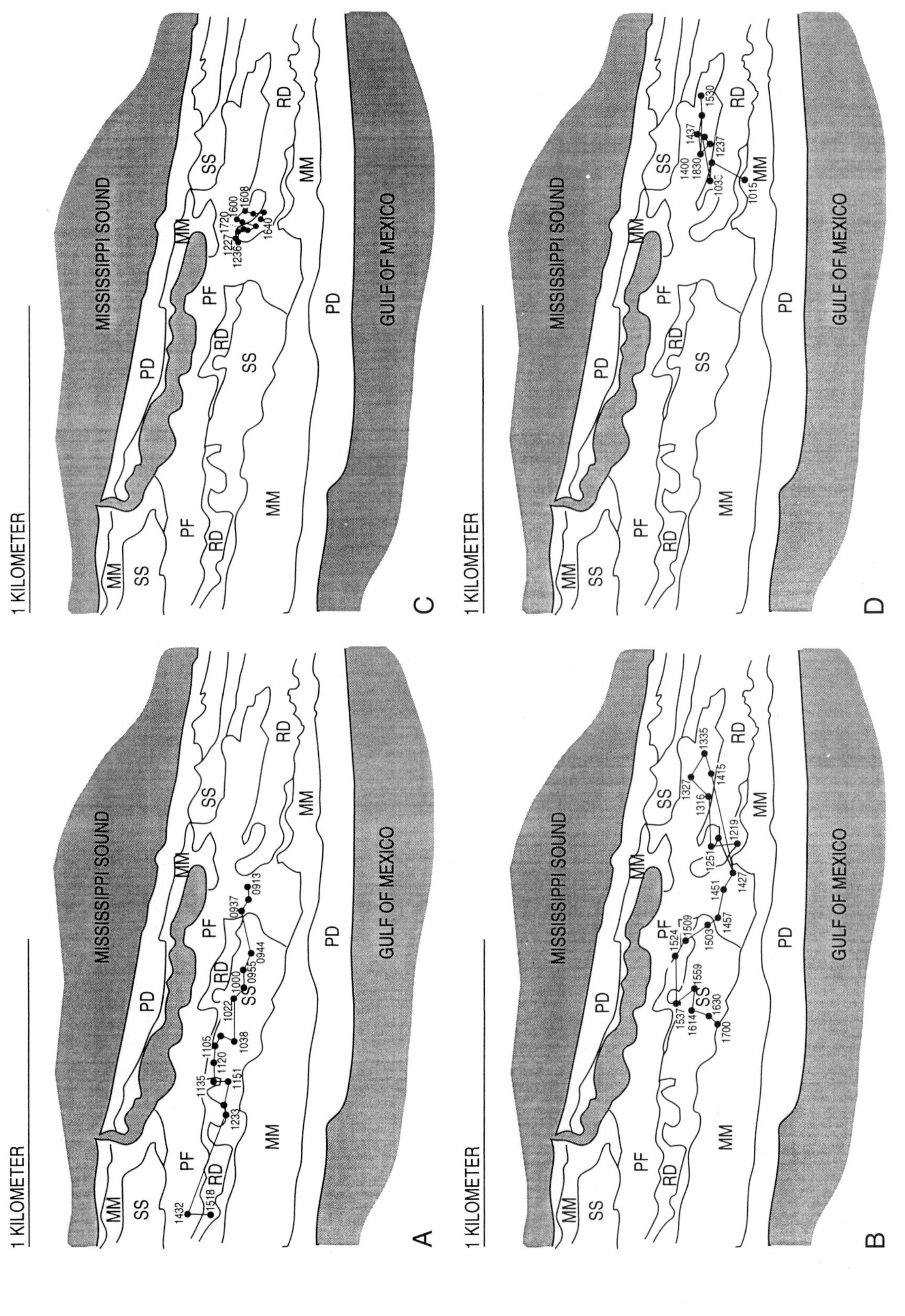
1 KILOMETER
MISSISSIPPI SOUND
GULF OF MEXICO
PD
PF
SS
RD
MM
A
B
C
D

EXPLORATION AND ENERGETIC CONDITION—AN EXAMPLE

The energetic status of a migrating bird is known to affect the likelihood of stopover (Moore and Kerlinger 1987, Kuenzi et al. 1991), length of stay (Rappole and Warner 1976, Yong and Moore 1993), foraging behavior during stopover (Loria and Moore 1991), and the response to the threat of predation (Moore 1994; D. Cimprich, unpubl. data). A migrant arriving at a stopover site in a fat-depleted condition is faced with the dilemma of needing to find quickly suitable habitat where it can deposit fat, while possibly not having the energy stores to adequately search among habitats. If high quality habitats are few in number or widely dispersed, fat-depleted migrants may be compelled to settle in a lower quality habitat, resulting in a longer stopover and delayed arrival on the breeding or wintering quarters.

Observations of neotropical landbird migrants that arrived along the northern coast of the Gulf of Mexico following a trans-Gulf flight suggest that birds may assess habitat during an initial exploratory phase (Moore et al. 1990, Aborn and Moore 1997). We used radio-telemetry to study the movement pattern of Summer Tanagers (*Piranga rubra*) during stopover following migration across the Gulf of Mexico (Moore and Aborn 1996, Aborn and Moore 1997).

Research was conducted on Horn Island (30° 14′ N, 88° 40′ W), a barrier island located approximately 20 km off the Mississippi coast. Vegetation on the island consists of a mosaic of five habitat types: pine forest, scrub/shrub, marsh/meadow, relic dune, and primary dune (see Moore et al. [1990a] for habitat descriptions). Birds were caught in mist-nets, fitted with radio transmitters (Custom Telemetry and Consulting, Inc., Watkinsville, GA; 1.3–1.4 g, 1-km range, 7-day life span), and tracked continuously until they left the island or the transmitter failed. We recorded the bird's location, habitat, and activity status (perched, active, or flying).

We calculated linear distance (distance between the point where bird was released to place it roosted at dusk), total distance moved, average distance per move, average time elapsed between each move, and average rate of movement. We also computed the angle of each turn and turn bias (i.e., whether the turn was left or right). An equal degree of left and right turns will result in an angular turn bias of 0°, whereas circling movement yields a larger turn bias, as if moving in a defined area or turning back on its previous move. Total distance divided by linear distance yields a "meander ratio" (Willamson and Gray 1975), which reflects area-restricted movement. We used vector analysis (Zar 1996) to determine whether a bird's sequence of moves were concentrated in a particular direction (r-statistic). Values close to zero indicate a high degree of variability in the directions (little directionality), whereas values that approach 1 suggest that birds tended to orient their movements in the same direction (high directionality). If Summer Tanagers move in a circular pattern or turn back on their previous movement the r-value should be small and statistically insignificant according to a Rayleigh test (Zar 1996).

Pattern of movement differed among individual Summer Tanagers that stopped over on Horn Island following trans-Gulf migration (Aborn and Moore 1997). Several factors undoubtedly contributed to observed variability, including the migrant's nutritional condition. Among the 24 Summer Tanagers tracked during spring migrations 1992 through 1994, 15 birds were classified as "fat" (body mass ≥15% above fat-free mass) and nine birds were "lean" (<15% above fat-free mass). Fat-free body mass equals 24.7 g for Summer Tanagers (Odum 1993). If we assume that the difference in body mass is fat and translate that difference into migratory flight range estimates (Pennycuick 1992), birds classified "fat" could fly on average 513 km flight distance, whereas lean birds could fly only 162 km.

Lean tanagers moved a longer linear distance, covered greater total distance, and moved at a faster rate than did fatter birds (Table 1; Fig. 1a,b). Fat tanagers were often perched for periods of an hour or more, whereas lean tanagers were rarely perched for extended periods (Table 1). Although the ratio of linear-to-total distance (meander ratio), which reflects the degree to which birds circumscribe their movements, did not differ between lean (ratio = 4.5) and fat (ratio = 5.0) birds, other measurements indicate that fat birds confined their movement to a smaller area than lean birds (Fig. 1c,d). Fat birds showed greater angular dispersion ($r = 0.248$) than lean birds ($r = 0.477$), which suggests that they did not concentrate their movements in a particular direction. Likewise, fat birds displayed a stronger turn bias (11.2) than lean birds (6.4), which is indicative of circling movement.

Habitat use also differed between the two groups (Fig. 2). Fat birds were located in pine forest twice as often as lean birds, whereas three-fourths of the radio locations for lean birds were in scrub/shrub habitat. Not only did habitat use differ between the two groups, but the tanagers used habitats out of proportion to habitat availability in different ways (Fig. 3), which suggests that they were actively selecting different habitats.

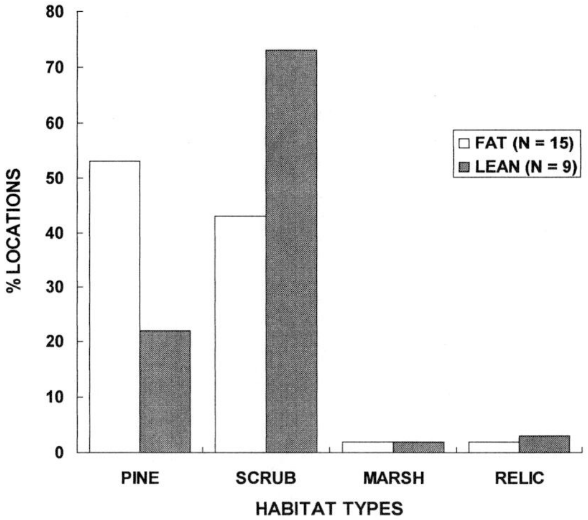

FIGURE 2. Habitat use by fat and lean Summer Tanagers radio-tracked on Horn Island, Mississippi, following migration across the Gulf of Mexico.

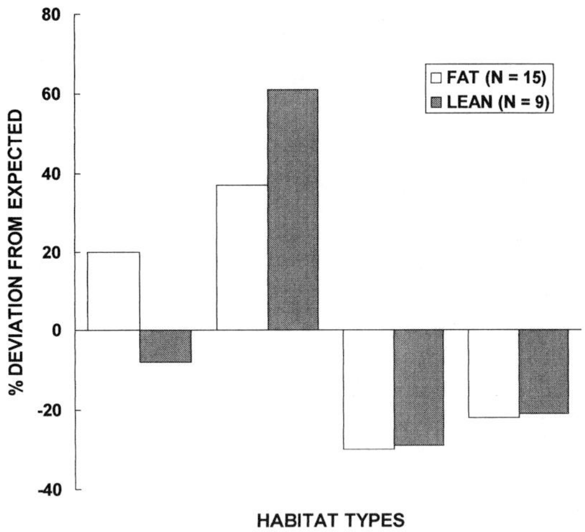

FIGURE 3. Deviation from expected use of habitat by fat and lean Summer Tanagers radio-tracked on Horn Island, Mississippi, following migration across the Gulf of Mexico. Expected habitat used is based on availability of habitat on the island.

What might account for the observed differences between fat and lean birds? Replenishment of depleted fat stores is probably a higher priority for lean than fat tanagers (see Loria and Moore 1990), and scrub/shrub habitat appears to be the most suitable place for doing so (F. R. Moore, unpubl. data). Previous work on Horn Island (Moore et al. 1990) showed a high degree of selectivity for scrub habitat by most migrants, despite its low availability, and many migrants that stop over on Horn Island have catabolized much of their fat stores while flying across the Gulf of Mexico.

In contrast, conservation of remaining fat stores and avoidance of predation are probably priorities for fat tanagers, which would explain less movement and a preference for pine forest. Observations of migrant landbird behavior in the presence of raptors suggest that pine habitat may offer greater concealment (D. A. Aborn and F. R. Moore, pers. obs.).

We would not be surprised if tanagers used different cues to assess habitat depending on their nutritional condition. Moreover, the value of information about habitat gained during stopover probably varies with the birds internal state such that lean birds may be more likely to pay costs associated with acquiring habitat information.

CONSTRAINTS ON *EN ROUTE* HABITAT SELECTION

Upon arriving at a stopover site, a migrant is faced with a mosaic of unfamiliar habitats. When animals find themselves in such settings, we would expect them to familiarize themselves with the kinds, distribution, and abundance of available resources and places safe from predators, to exploit resources efficiently, and to replenish energy stores quickly (cf. Orians and Wittenberger 1991; see Hutto 1985b). Yet, time available for searching is an important component of any habitat selection process (Ward 1987), and migrants are probably time-constrained during passage (sensu Alerstam and Lindström 1990). The time constraints imposed on a migrating bird probably intensify the conflict between the value of information gained through exploration and pressure to minimize time spend *en route.*

How might the constraint of time affect the process of habitat selection? The threshold for acceptance of habitat during migration depends on the migrant's time program (sensu Berthold 1990; Gwinner 1986) and internal state (e.g., fat stores), which may be modified by external factors (e.g., competitors, weather). It may be that observed distributions of migrants among habitats reflect a lowered threshold of acceptance or errors in habitat assessment. When time for habitat assessment is brief, we expect migrants to obtain information on the quality of different habitats using cues that are virtually instantly assessable. We also expect migrants to use simple proximate cues for habitat choice, because a simple cue is more likely to occur in a variety of habitats (cf. Morton 1990). Moreover, we would expect migrants to be selective at first, but as time passes, to select less suitable habitats (Ward 1987), which means that the threshold for acceptance decreases over time.

Whereas evidence leads us to expect more exploration as circumstances become less familiar (Bell 1991), not to mention more exploration as the area becomes less suitable, time-constrained migrants may not have the time nor the energy for extensive search, especially if the area is of poor quality. Rapid exploration that yields incomplete information about the distribution of resources and sources of stress would not be surprising during stopover. The few studies that have examined the distributions of captured migrants among different habitats during stopover (Bairlein 1983, Moore et al. 1990, Winker 1995) suggest that exploration prior to settlement takes place quickly, probably within an hour of arrival, if at all. It may simply be best for birds that can not gain access to suitable habitats or experience difficulty distinguishing habitats, to continue migration (Rappole and Warner 1976, Terrill 1988).

Searching efficiency is an important component of the habitat selection process, which leads us to expect age and migratory experience to influence *en route* habitat selection. Distribution of migrants among habitats in the Alps (Bruderer and Jenni 1988) and along riparian corridors of the Rio Grande River (Yong et al. 1998) suggests that juveniles are more likely to land in unfavorable habitats than adults. Bairlein (1983) attributed age-specific differences in habitat use among European migrants to inaccuracies in habitat preference in young birds. *En route* "problems" are undoubtedly magnified for hatching-year birds on their first migration (Ralph 1978, Alerstam 1978, Gauthreaux 1982a, DeSante 1983, Moore 1984, Lindström and Alerstam 1986, Terrill 1988, Woodrey and Moore 1997, Woodrey *this volume*), and individuals with different levels of migratory experience can be expected to respond differently to the exigencies of migration. Hatching-year birds may be more likely to land in unfavorable habitats than adults and once landed, they may lack the experience to efficiently search an area. Hatching-year migrants are also often behaviorally subordinate to adults (Terrill 1988), which could affect their reaction to other migrants *vis-a-vis* habitat assessment. In addition, experience is likely to effect assessment of habitat in relation to conflicting demands between energetics and other environmental factors (Yong et al. 1998). Choice of feeding location *vis-a-vis* predation risk and energetic requirements, for example, may be age-dependent (Cresswell 1994).

Mechanisms of habitat selection may also vary with the migrant's sex to the extent that habitat use is sex-specific during passage (e.g., Yong et al. 1998). Outside of the migratory period, males and females may settle in different habitats (cf. Lynch et al. 1985, Parrish and Sherry 1994) or use the same habitat differently (Morse 1989), and Morton (1990) found that male and female Hooded Warblers (*Wilsonia citrina*) use different structural cues to chose winter habitat. Laboratory experiments designed to study habitat segregation between sexes in Hooded Warblers found sex-specific preferences tied to vegetation structure consistent with differential habitat use on the wintering grounds (Morton 1990; see also Ornat and Greenberg 1990). That females and males might use different habitats or the same habitat differently warrants attention. For example, nutritional demands in relation to breeding performance differ between sexes. If efforts to satisfy differential demands take place during passage, sex-specific use of habitat may occur. Furthermore, if social dominance (Carpenter et al. 1993a,b; Parrish and Sherry 1994, Marra et al. 1993) manifests itself during passage, females may occupy different habitats by virtue of their status.

ACKNOWLEDGMENTS

We thank P. Kerlinger, J. Busby-High, D. Cimprich, J. Clark, M. Guilfoyle, S. Mabey, M. Woodrey, and W. Yong for their helpful comment during preparation of the manuscript. This research was funded by NSF grants BSR-9020530 and BSR-9100054 to F. R. Moore, and a Frank M. Chapman Memorial Fund grant to D. A. Aborn.

Studies in Avian Biology No. 20:43–52, 2000.

AGE-DEPENDENT ASPECTS OF STOPOVER BIOLOGY OF PASSERINE MIGRANTS

MARK S. WOODREY

Abstract. Most studies of the behavior and ecology of birds address questions at the species level without assessing the effects of factors such as age, sex, or age-sex classes, although variation observed within species can be at least partly attributed to these factors. My objective is to synthesize known information concerning aspects of age-dependent *en route* ecology and behavior, emphasizing long-distance, intercontinental passerine migrants. Adults and yearlings show differences in the amount of fat stores carried during migration, in food acquisition rates in some species, and in the effect of differential social dominance on foraging sites used during migration. Foraging sites influence predation risk incurred by individuals, and young birds typically vary more in their ability to orient during migration than experienced birds. Results presented here emphasize the need to consider intra-specific, age-dependent effects on the behavior and ecology of migratory birds.

Key Words: age-dependent, competition, energetics, foraging, migrants, migration, orientation, passerine, predation, social dominance, stopover.

Migration is an ecological strategy evolved in response to temporal and spatial environmental heterogeneity that allows organisms to exploit seasonal resources (Baker 1978, Dingle 1980). For example, nearctic-neotropical (Hayes 1995) migratory landbirds leave their breeding sites in temperate regions before their food resources become scarce, travel thousands of kilometers in uncertain weather over ecological barriers, stop periodically to (re)build fat deposits, and finally arrive in tropical wintering habitats different from those in the temperate zone. After residing for five or six months in tropical communities, they return north again to their temperate breeding areas.

Despite renewed interest in the biology and conservation of long-distance passerine migrants (see Keast and Morton 1980, Hagan and Johnston 1992, Crick and Jones 1992, Finch and Stangel 1993, Martin and Finch 1995, and references therein), relatively few aspects of the stopover biology of passerine birds have been examined. Topics such as migrants and their relation to stopover habitats (Bairlein 1992a, Winker et al. 1992a), effects of habitat suitability on migrant stopover biology (Moore and Simons 1992a, Winker et al. 1992a), species-specific migration strategies (Bairlein 1992b), genetic control of migration (Berthold and Helbig 1992), energetic condition and its consequences (Biebach et al. 1986, Biebach 1992, Woodrey and Moore 1997), and conservation and management of migratory birds (Wood 1992, Moore and Woodrey 1993; Moore et al. 1993, 1995) have been investigated. Unfortunately, few studies have addressed age-specific aspects of migrant ecology and the possible consequences for a successful migration.

RATIONALE FOR FOCUS ON AGE-DEPENDENT STOPOVER BIOLOGY

The rationale for a focus on age-dependent stopover biology is two-fold. First, a migrant's fitness is enhanced by surviving in the best possible condition to maximize the probability of a successful migration. A successful migration is dependent upon how effectively the bird offsets the costs of migration (i.e., satisfies energy demand and meets *en route* contingencies; Alerstam and Lindström 1990). Regardless of whether benefits of migration accrue through increased productivity, increased probability of surviving an unfavorable season, or both, they must be balanced against increased mortality experienced during migration (e.g., Fretwell 1980, Alerstam and Högstedt 1982). The mortality associated with intercontinental migration, though difficult to estimate, is probably substantial (Lack 1946, Moreau 1972, Ketterson and Nolan 1982), and because this cost (reduced fitness, increased mortality) is absorbed largely by the hatching-year age-class (Nisbet and Medway 1972, Johnson 1973, Alerstam 1978, Greenberg 1980, Goss-Custard and Dit Durell 1983; Ketterson and Nolan 1982, 1983, 1985; DeSante 1983, Ramos 1988), differential costs should be reflected by age-dependent differences in stopover ecology.

Second, individuals with different levels of migratory experience can be expected to respond differently to the exigencies of migration (Ketterson and Nolan 1985, 1988; Terrill 1988). In addition to the energetic costs of transport (see Pennycuick 1975), migrants must (a) deal with unfamiliar habitats, which often differ in suitability (Bibby et al. 1976, Mehlum 1983, Moore and Simons 1992a), (b) resolve the conflicting demands of predator avoidance, food acquisition, and timely arrival on the breeding and wintering grounds (Metcalfe and Furness 1984, Lindström 1989, Moore et al. 1990, Moore 1994), (c) compete with other migrants and resident birds for limiting resources (Rappole and Warner 1976, Alerstam 1978, Bibby and Green 1980, Lindström and Alerstam 1986, Hansson and Pettersson 1989, Moore and Yong 1991), (d) respond to adverse weather conditions (e.g., Richardson 1978, 1990), and (e) correct for orientation errors (Ralph 1978; Moore 1984, 1990a; Alerstam 1990). Moreover, favorable *en route* habitat, where energy stores can be rapidly accumulated, is probably limited for migrants (Martin 1985, Martin and Karr 1986), or effectively so because migrants do not have the luxury of searching for the "best" stopover site (see Hutto 1985a, Moore et al. 1993). These problems are presumably magnified for hatching-year birds on their first migration due to their lack of experience (Ralph 1978, Alerstam 1978, Goss-Custard and Dit Durell 1983, Gauthreaux 1982a, DeSante 1983, Moore 1984, Lindström and Alerstam 1986, Terrill 1987) and socially subordinate status (Gauthreaux 1978, Terrill 1987, Marchetti and Price 1989, Sherry and Holmes 1989, Woodrey 1995).

My purpose is, first, to synthesize and evaluate current knowledge about age-dependent aspects of migration, and, second, to identify areas of stopover biology where age-dependent information is lacking. The focus of this review is primarily on passerine migrants because of the growing interest in this group of birds in North America and Europe (see Keast and Morton 1980, Hagan and Johnston 1992, Crick and Jones 1992, Finch and Stangel 1993, Martin and Finch 1995, and references therein). This assessment of age-dependent stopover biology begins by addressing differences in the timing of migration between age classes, then discusses *en route* challenges facing migratory birds, including energetics, food acquisition, competition and social dominance, predation, and orientation. I conclude with a summary of the significance of age-dependent research and provide direction for future research.

AGE-DEPENDENT DIFFERENCES IN TIMING AND LOCATION OF MIGRATION

Although differential timing of migration by age has been documented, the proximate causes for differences have not been thoroughly investigated and interpreted. Early studies showed that in many cases hatching-year (HY: yearling) individuals were the first to leave their natal territories because after-hatching-year (AHY: adult) individuals were delayed until they completed their prebasic molt (reviewed by Gauthreaux 1982b). Numerous studies indicate that during fall migration adult passerines precede immatures (Hussell et al. 1967, Ely 1970, Leberman and Clench 1973, Johnson 1974, Seel 1977; Hussell 1980, 1981, 1982, 1991; Woodrey and Chandler 1997), while others indicate that immatures precede adults (Magee 1924, cited in Gauthreaux 1982a; Leberman and Clench 1973, Woodrey and Chandler 1997), while yet others found no age difference in timing (Murray 1966, Woodrey and Chandler 1997). Thus, no consistent pattern emerges within or between groups of species in age-specific timing of migration. Note however, that the lack of temporal overlap is important when discussing the role of intraspecific competition on the behavior of migrants during stopover (see COMPETITION AND SOCIAL DOMINANCE below).

Age-classes of passerines often migrate along different routes, with juveniles concentrated along coastal areas. During fall migration, young birds comprise 85–95% of the birds captured along the Atlantic coast (Drury and Keith 1962, Murray 1966, Ralph 1981, Morris et al. 1996) and Pacific coast (Ralph 1971, Stewart et al. 1974). In contrast, only 65–70% of the birds captured at inland sites are young birds (Stewart et al. 1974, Ralph 1981). This "coastal effect" might be the result of different migratory routes followed by young and adults (Leberman and Clench 1975), or the high percentages of young could denote the periphery of a species' migration route (Ralph 1981). Four of six species studied by Woodrey and Moore (1997) along the northern coast of the Gulf of Mexico showed more balanced age-ratios (69.7% young birds). Species studied by Ralph (1981) follow more restricted routes and generally move parallel to the coast line, whereas birds migrating through the central portion of the United States display a more broad-front migration (i.e., less geographically restricted) than coastal migrants. Furthermore, birds moving south in the fall are traveling perpendicular to the east-west orientation of the northern Gulf coastline. Thus, more adult and juvenile birds encounter the Gulf coast, resulting in more balanced age-ratios.

EN ROUTE CHALLENGES

The variety of problems or contingencies faced by migratory landbirds during the *en route* period of their annual cycle may be classified, somewhat arbitrarily, into five topics: (a) the energetic cost of transport, (b) food acquisition, (c) competition and social dominance, (d) predator avoidance, and (e) orientation.

ENERGETIC COST OF TRANSPORT

Before initiating their journey, migratory birds add an average of 30–50% of lean body mass in fat stores (Blem 1980). During migration, free-ranging birds are capable of replenishing depleted fat stores at rates approaching 10% of body mass/day (Dolnik and Blyumental 1967, Bairlein 1985a, Biebach et al. 1986, Moore and Kerlinger 1987). Fat stores become important when migrants face the heightened energy demand of making long-distance, non-stop flights over ecological barriers such as a large body of water or a desert (Biebach et al. 1986, Moore and Kerlinger 1991).

Visible subcutaneous fat stores are often used as a surrogate measure of fitness in migrating birds because it is difficult, if not impossible, to measure the direct effects of *en route* events on survival or reproductive success. Despite changes in mass due to changes in the flight musculature (Marsh 1984, Lindström and Piersma 1993), I assume here that fat accounts for essentially all mass differences during stopover (Odum et al. 1961, 1964; Rogers and Odum 1966, Child 1969). In addition, the interest here is comparisons of arrival masses between age classes within species, and I assume that any other components contributing to differences in mass is the same across age classes.

Of the 26 cases (21 species) for which age-specific stopover data are available, adults were significantly heavier than young for 13 cases (11 species; Table 1). Inexperience and subordinate social status are implicated as potential causes of greater arrival mass of adult American Redstarts (see Table 1 for scientific names), Swainson's Thrushes, and White-eyed Vireos (Woodrey and Moore 1997), and Bluethroats (Ellegren 1991). These factors also likely influence age-dependent body mass differences in Blackpoll and Yellow-rumped warblers (Murray 1979), Pied Flycatchers (Viega 1986), Sedge Warblers (Spina and Bezzi 1990, Basciutti et al. 1997), Traill's Flycatchers, Red-eyed Vireos, Tennessee Warblers, American Redstarts, and Northern Waterthrushes (Morris et al. 1996).

Yearling migrants are likely at a competitive disadvantage, as mentioned previously, because of their inexperience and socially subordinate status relative to adults. Younger, less experienced birds are typically less efficient at procuring resources (Burger 1988, Wunderle 1991), and are usually subordinate to adults during the nonbreeding season (Terrill 1987, Sherry and Holmes 1989). Social status could handicap yearling migrants if their subordinate status affects access to resources and the subsequent opportunity to deposit necessary fat reserves (cf. Ekman and Askenmo 1984, Piper and Wiley 1990).

Age-dependent differences in body mass likely have profound consequences for migrant populations. Although the effects of fat stores on survival and reproduction are difficult to measure directly, we can speculate as to their effects. For instance, Woodrey and Moore (1997) calculated flight range estimates by age-class for each of the six species they studied. In a species where adults were significantly heavier than young birds upon arrival along the northern coast of the Gulf of Mexico, adult American Redstarts with an average mass of 9.4 g can fly 1,400 km, whereas an average immature bird has a 950 km flight range (average mass = 8.4 g). Approximately 47% of immature American Redstarts have inadequate stores to complete a successful trans-Gulf crossing, which could result in death due to starvation during the 1,200 km non-stop flight over the Gulf of Mexico.

In addition to starvation, there may be more subtle and less dramatic consequences of age-specific fat deposits. Young migrants in a fat depleted state have a smaller margin of safety to buffer the effect of adverse weather on the availability of insect prey during stopover (sensu Moore and Kerlinger 1991). Also, the need of young birds to satisfy energetic demands of migration through increased foraging (equals increased activity) may expose them to increased predation relative to adults (Metcalfe and Furness 1984, Moore 1994, Woodrey 1995). Further, because of increased duration of stopover, as required to replenish depleted stores, the migration of young birds may be slowed, which may jeopardize opportunities to secure suitable winter territories (Stutchbury 1994).

Differing flight morphologies are known to affect the aerodynamics of migratory flight. Long-distance migrants typically have wings with a high aspect ratio and low loading, which reduces drag and the energetic cost of powered flight (Pennycuick 1975; Rayner 1988, 1990). Differences in wing lengths of adult and juvenile birds within a species have been noted for a variety of species. Almost without exception, young passerine birds have shorter wings than adult birds (Alatalo et al. 1984). Thus, given the same mass and wing shape, young birds have a shorter flight range compared to adults because of the greater mechanical efficiency of flight and lower cost of transport in longer-winged birds (Rayner 1990). Shorter wings, in combination with lower fat stores for migration, further decrease the probability of a successful migration for young migratory passerines.

Unlike species discussed previously, yearling Gray Catbirds, Magnolia Warblers (Woodrey and Moore 1997), and Chaffinches (Å. Lindström, unpubl. data) were significantly heavier than adults upon arrival at a particular stopover site (Table 1). This unexpected result for Gray

TABLE 1. SUMMARY OF SELECTED RESULTS FROM STUDIES OF AGE-SPECIFIC STOPOVER BIOLOGY OF LONG-DISTANCE PASSERINE LANDBIRDS

Species	Fat stores	Body mass	Length of stopover	Mass change[a]	Reference
Eastern Wood-Pewee *Contopus virens*	NS[b]	NS	(?)[c]	NS	Morris et al. 1996
Traill's Flycatcher *Empidonax traillii*	NS	AHY>HY[d]	(?)	NS	Morris et al. 1996
Yellow-bellied Flycatcher *E. flaviventris*	NS	NS	(?)	NS	Morris et al. 1996
Bluethroat *Luscinia svecica*	AHY>HY	(?)	HY>AHY	NS	Ellegren 1991
Veery *Catharus fuscescens*	NS	NS	NS	NS	Morris et al. 1996
Swainson's Thrush	NS	NS	(?)	NS	Morris et al. 1996
Catharus ustulatus	AHY>HY	AHY>HY	(?)	NS	Woodrey and Moore 1997
Sedge Warbler	AHY>HY	AHY>HY	(?)	(?)	Spina and Bezzi 1990
Acrocephalus schoenobaenus	AHY>HY	AHY>HY	(?)	(?)	Basciutti et al. 1997
Pied Flycatcher *Ficedula hypoleuca*	(?)	AHY>HY	HY>AHY	NS	Veiga 1986
Gray Catbird *Dumetella carolinensis*	NS	HY>AHY	NS	NS	Woodrey and Moore 1997
White-eyed Vireo *Vireo griseus*	AHY>HY	AHY>HY	NS	(?)	Woodrey and Moore 1997
Red-eyed Vireo	NS	AHY>HY	NS	NS	Morris et al. 1996
Vireo olivaceus	NS	NS	(?)	(?)	Woodrey and Moore 1997
Tennessee Warbler *Vermivora peregrina*	AHY>HY	AHY>HY	NS	NS	Morris et al. 1996
Cape May Warbler *Dendroica tigrina*	NS	NS	NS	NS	Morris et al. 1996
Magnolia Warbler	NS	NS	(?)	NS	Morris et al. 1996
Dendroica magnolia	NS	HY>AHY	(?)	(?)	Woodrey and Moore 1997
Yellow-rumped Warbler *Dendroica coronata*	(?)	AHY>HY	(?)	(?)	Murray 1979
Bay-breasted Warbler *Dendroica castanea*	NS	NS	NS	NS	Morris et al. 1996
Blackpoll Warbler	NS	NS	NS	NS	Morris et al. 1996
Dendroica striata	(?)	AHY>HY	(?)	(?)	Murray 1979
Black-and-white Warbler *Mniotilta varia*	NS	NS	(?)	NS	Morris et al. 1996
American Redstart	AHY>HY	AHY>HY	NS	NS	Morris et al. 1996
Setophaga ruticilla	AHY>HY	AHY>HY	NS	NS	Woodrey and Moore 1997
Northern Waterthrush *Seiurus noveboracensis*	AHY>HY	AHY>HY	HY>AHY	NS	Morris et al. 1996
Wilson's Warbler *Wilsonia pusilla*	AHY>HY	(?)	HY>AHY	NS	Yong et al. 1998
Chaffinch *Fringilla coelebs*	HY>AHY	HY>AHY	(?)	(?)	Å. Lindström, unpubl. data

[a] Based on data from recaptured birds.
[b] NS = non-significant difference between adult and yearling birds.
[c] (?) indicates that data were insufficient to draw biological or statistical conclusions, results were not reported, age comparisons not made, or variable not measured as part of study.
[d] AHY = after-hatching-year bird (adult); HY = hatching-year bird (yearling); AHY>HY or HY>AHY = significant difference between adult and young birds.

Catbirds may be related to its migration strategy, which is to migrate around, rather than across, the Gulf of Mexico (Eddins and Rogers 1992). Furthermore, Christmas Bird Count data show that Gray Catbirds winter abundantly from Florida to southern Texas (Root 1988). This tendency for many catbird individuals not to cross the Gulf of Mexico may explain why few individuals arrive with sufficient fuel to complete a trans-Gulf crossing (Woodrey and Moore 1997). Nevertheless, age differences in fat stores could be important if catbirds exhibit age-dependent habitat or geographic segregation in winter (e.g., Ketterson and Nolan 1983). Apparently, yearling Magnolia Warblers and Chaffinches compensate in some way for their lack of experience and their socially subordinate status so as to maintain greater fat stores than adults. At present we do not understand the mechanism for such high yearling fat stores.

Food Acquisition

Acquisition of adequate food resources to meet the anticipated energy demand of migration is possibly the most important constraint on migratory birds during their annual cycle. Two factors that may influence the foraging behavior of passerine birds are experience and social dominance (Gauthreaux 1978, Burger 1988, Wunderle 1991; see Competition and Social Dominance below). Experience can affect how, when, and where a bird might forage. Many studies concerning age-related (i.e., experience) foraging behavior focus on the structure of the foraging bout (e.g., Ziegler 1976, Lovette and Holmes 1995). A change in feeding bout structure, caused by changes in, for example, search time, handling time, inter-food interval, or bout length could lead to a change in the food intake rate for an individual.

When food is plentiful, even a less efficient forager may have few problems finding enough food to deposit sufficient lipid stores. However, any decrease in the foraging efficiency (e.g., foraging rate) could be critical to a juvenile migrant about to cross an ecological barrier because experience and/or social dominance may affect the opportunity to secure adequate resources for fat deposition (Ekman and Askenmo 1984, Lindström et al. 1990). Because yearling migrants are inexperienced and often socially subordinate to adults, they might be expected to satisfy the energy requirements and meet *en route* contingencies less effectively than adults (e.g., HY migrants deposit fat at a slower rate than AHY migrants).

Only four studies provide any empirical evidence relevant to the expectation of slower rates of fat deposition in yearling migrants (Table 1). Fat deposition rates for adult and juvenile Bluethroats migrating through Sweden were almost identical (adults = 0.089 g/d, juveniles = 0.092 g/d; Ellegren 1991). However, adults put on more fat than juveniles in a given stopover period because adults were more likely not to lose mass at the beginning of the stopover period. During stopover, young Bluethroats lost mass initially and then began gaining mass on the third day of the stopover period. Adult Bluethroats tended not to lose mass at any time, and they showed consistently higher rates of mass gain for any particular day (Fig. 1), although this was not statistically significant. Fat deposition data, based on recaptured individuals, showed no significant difference between adult

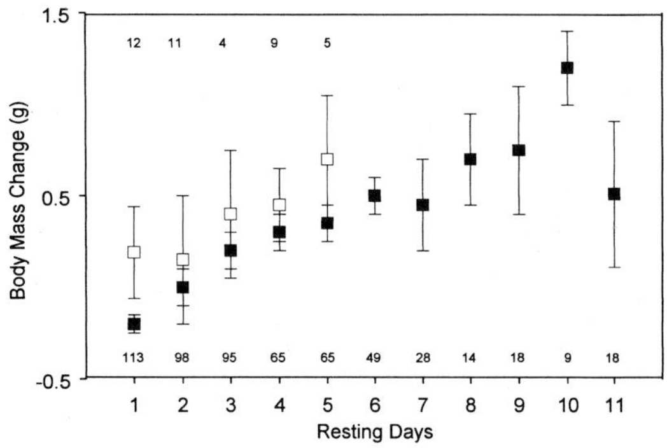

FIGURE 1. Pattern of fat deposition of fall migrating Bluethroats indicated by the average change in body mass as a function of the number of days from banding to recapture. The vertical lines represent ± 1 SE. The open squares are adults, the filled squares immatures. Sample sizes are represented by numbers above (adults) and below (juveniles) (from Ellegren 1991).

and yearling Wilson's Warblers (Yong et al. 1998), American Redstarts and White-eyed Vireos (Woodrey and Moore 1997), Pied Flycatchers (Veiga 1986), nor any of the 14 species studied by Morris et al. (1996; Table 1).

Because few migrants are recaptured during migration (Veiga 1986, Ellegren 1991; Winker et al. 1992a,b; Morris et al. 1996, Woodrey and Moore 1997), Winker et al. (1992b) used linear regression to examine species-specific relationships between body mass and time of day of capture for individuals captured only once. Using this approach to investigate age-specific rates of fat deposition, Woodrey and Moore (1997) found that yearling White-eyed Vireos and American Redstarts showed a significant positive relationship between body mass and time of day whereas adults showed no significant relationship. These contradictory results, in conjunction with the fact that for this analysis no study showed a significant age difference in mass change, suggest that the expectation of greater rates of fat deposition for yearlings may not be valid. Although Woodrey and Moore (1997) found a statistically significant relationship for yearlings, the mass gained per day resulted in little increase in estimated flight ranges for these migrants (White-eyed Vireo = 4.68 km, American Redstart = 7.30 km). The results from these studies are clearly ambiguous with regards to rates of fat deposition for different age classes of migrant passerines.

Few studies have dealt with foraging behavior of passerines during migration and even fewer have looked at age-related differences during stopover. However, Woodrey (1995) compared the foraging behavior of HY and AHY American Redstarts during stopover. Both adult and yearling redstarts frequently moved forward while foraging, though they sometimes made acute turns to the side or to the rear. The angular movements of yearling and adults did not differ significantly, nor did the rate or speed at which they foraged. Young redstarts did show greater variation in the rate and speed at which they foraged.

One possible explanation for the lack of observed differences in the foraging rates and patterns of adult and immature redstarts may be related to experience and learning in juvenile birds. Whenever the foraging ability of adult and full-grown juvenile birds have been compared, the former typically have greater success (Burger 1988, Marchetti and Price 1989, Wunderle 1991). However, it is possible that an age-related disparity in foraging success may disappear by the onset of migration (Sullivan 1988, Weathers and Sullivan 1991). In their discussion of developmental constraints on age-related foraging differences in birds, Marchetti and Price (1989) concluded that young birds appear to be under strong selection to reach adult form and function as rapidly as possible. Thus, any age-related disparity in foraging efficiency of migrant passerines could disappear before they begin their migration.

Another aspect of foraging with respect to age-classes of birds is neophobia (Greenberg 1983, 1984a,b,c, 1990). Neophobia occurs when a bird responds to novel situations with acute stress, which leads to avoidance (Greenberg 1990). If young birds are less likely to approach novel prey items or novel microhabitats, then fewer opportunities will be available for HY foraging repertoires, resulting in reduced ecological plasticity. Neophobia in juvenile birds could play an important role in shaping the use of microhabitat sites and the exploitation of unfamiliar habitats during migration, resulting in young birds being less efficient at satisfying the energetic demands of migration.

Competition and Social Dominance

One prerequisite for competition is that potential competitors, for example different age classes, should overlap temporally. As noted previously however, many species exhibit age-specific differential timing of passage (see AGE-DEPENDENT DIFFERENCES IN TIMING AND LOCATION OF MIGRATION above). However, despite age differences in the mean passage times of birds, yearlings and adults often occupy the same stopover site simultaneously (e.g., Woodrey and Chandler 1997, Woodrey and Moore 1997), increasing the potential for competitive interactions among migrants. Competition for food resources may reduce fat deposition rates during stopover, particularly considering the energetic demand of migration (Moore and Yong 1991). In addition, migrants concentrated in unfamiliar habitats where resources are limited must contend with other individuals that have similar dietary requirements. Some evidence, albeit mostly circumstantial, is consistent with the occurrence of food-based competition during migration (reviewed by Moore and Yong 1991): (1) observations of territoriality among migrants, (2) density-dependent settlement during migration, and (3) habitat selectivity in relation to food availability. Moore and Yong (1991) provide direct evidence for competition with a predator-exclosure experiment, which indicates that passerine migrants depress food abundance during stopover, and that migrants replenish fat stores at rates inversely proportional to migrant densities.

What are the possible effects of food-based competition in combination with the differing dominance relationships among age-classes of passerine migrants? Social dominance, which may affect an individuals foraging behavior, frequently confers priority of access to resources (Gauthreaux 1978; see also Piper 1997 for review of social dominance in birds). Individuals may have a different social status depending on characteristics such as sex, age, size, and aggressiveness (Thompson 1960; Brown 1963, 1975; Waite 1987). Many studies of nonbreeding birds have documented differences among age and sex classes in resource use (reviews by Selander 1966, Morse 1980a, Grubb and Woodrey 1990, Wunderle 1991), and the mechanism of interference competition through social dominance is often implicated in differences in resource use. To compensate for their lower social status, subordinate individuals have two choices: use non-preferred resources or adopt other compensatory behaviors. In an observational study of free-ranging American Redstarts during stopover along the northern Gulf coast, Woodrey (1995) found that age-classes of foraging American Redstarts differed significantly in use of trees: HY birds used slash pines (*Pinus elliottii*), adults used sand live oaks (*Quercus geminata*; Woodrey 1995). Hatching-year birds gleaned insects from needle clusters located near the outer portions of the branches, up to a relative height in trees of 3/4, whereas adults typically hawked insects from the middle areas of the lower branches of tall pine trees. These observations led Woodrey (1995) to hypothesize that social dominance is the mechanism responsible for the differential use of microhabitats by adult and juvenile redstarts. In addition, Woodrey (1995) showed that social interactions (intraspecific chases and supplanting attacks) of American Redstarts during stopover along the northern coast of the Gulf of Mexico were correlated with the density of conspecifics. The interactions were facilitated by species-specific agonistic displays, including wingspread and tail-fanning (Ficken 1962, Ficken and Ficken 1962, Holmes et al. 1989), and individuals vocalized with contact call notes throughout the day. In combination, these observations support the hypothesis that social dominance is the mechanism whereby adult American Redstarts meet the contingencies of migration more effectively than yearlings. Thus, social status does appear to affect a yearling redstarts ability to deposit necessary energy stores (e.g., Ekman and Askenmo 1984, Lindström et al. 1990) and it may expose young individuals to increased predation risk (see Piper 1997:167; see also PREDATOR AVOIDANCE below).

The role of intraspecific interference competition in shaping migration strategies is illustrated by studies of Rufous Hummingbirds (*Selasphorus rufus*; Carpenter et al. 1993a,b). Three recognizable age-sex classes migrating through the Sierra Nevada Mountains of California overlapped temporally and defended feeding territories (Carpenter et al. 1993a). These classes differed in their ability to acquire and defend high quality feeding territories, and these differences affected resource use. For example, immature males defended the highest quality patches, whereas young females defended the poorest (Fig. 2). Thus, one would expect females to be less efficient at satisfying the demands of migration. However, immature female hummingbirds gained body mass at the same rate as immature males and adult females. Moreover, departure body masses for females were not significantly different from male hummingbirds (Carpenter et al. 1993b). Carpenter and her colleagues suggest that females may be energetically compensated by (1) lower costs of flight incurred during foraging and defense, a result of their lower wing disc loading, and (2) greater success at robbing nectar from male's energy-rich territories, likely facilitated by duller plumage coloration and, in the case of adult females, greater experience.

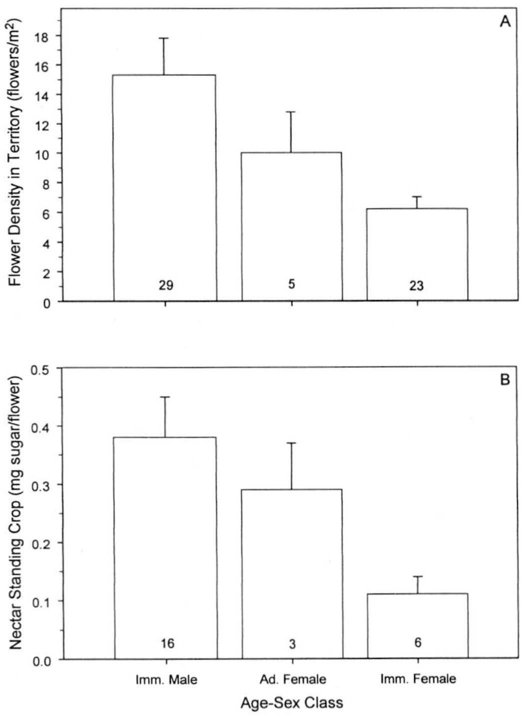

FIGURE 2. Variation in measures of territory quality among different age-sex classes of Rufous Hummingbirds. Means and standard errors are shown, with sample sizes inside the bars. A. Average flower densities in territories pooled across all years (from Carpenter et al. 1993a). B. Nectar standing crop measured on territories in the evening (1930–2030 hrs Pacific Daylight Time) in 1980 (from Carpenter et al. 1993a).

PREDATOR AVOIDANCE

Predation pressure on migrating birds can be intense (Rudebeck 1950, 1951; Walter 1979, Kerlinger 1989, Lindström 1990a, Moore et al. 1990, Aborn 1994). Predation can be particularly severe at stopover sites because (1) migrants may find themselves concentrated in unfamiliar habitats, and (2) the passage and concentration of avian predators has been shown to be correlated with the passage and concentrations of migrants (Kerlinger 1989, Aborn 1994). Predation pressure has been shown to affect the rates of food intake of migratory birds (Metcalfe and Furness 1984) and their habitat selection (Lindström 1989). Migratory birds about to embark on a long distance flight must balance the conflicting demands of food acquisition and predator avoidance during stopover (Metcalfe and Furness 1984, Moore 1994). Decisions concerning the trade-off between food intake and predation risk depend in part upon both the risk of predation and the cost of lost foraging opportunities (Ydenberg and Dill 1986, cited by Moore 1994). Hatching-year birds face two problems relative to after-hatching-year birds when making this trade-off. First, because HY birds are significantly leaner than adults, the former may have to resume foraging more quickly than adults, thus exposing themselves to increased predation risk (Moore 1994). Second, the subordinate status of HY birds may further expose them to an increased risk of predation because dominant individuals often force subordinates into riskier foraging sites through interference (Ekman and Askenmo 1984, Koivula et al. 1994, Woodrey 1995; see also Piper 1997:167). Census data of avian predators migrating along Ft. Morgan peninsula in coastal Alabama indicate that bird-eating raptors are found disproportionately in pine-shrub habitats (D. Cimprich and F. Moore, unpubl. data), the same habitat where young, socially subordinate redstarts are found foraging on the outer portions of slash pines (Woodrey 1995). The combination of these observations suggest that young birds are exposed to increased predation risk relative to adults during stopover.

Orientation

Numerous studies have shown that experience affects the orientation behavior of migrant passerines (see Gauthreaux 1982a, Moore 1984, Sandberg et al. 1991, and references therein). The majority of studies show that naive migrants consistently display greater angular deviations in orientation when compared to older, more experienced adult birds (Drury and Keith 1962; DeSante 1973, 1983; Able 1977; Ralph 1978, 1981; McLaren 1981, Moore 1984; Sandberg et al. 1988, 1991). For example, adult Savannah Sparrows (*Passerculus sandwichensis*) showed angular deviations half the magnitude of immatures when placed into orientation cages and their nocturnal orientation direction was observed (Moore 1984; Fig. 3). Increased variation in orientation by immature birds means more errors, which will increase the risk of inadequate fat stores for long, non-stop migratory flights (e.g., Woodrey and Moore 1997). Even if the bird survived the extended flight, it would likely be in an energetically depleted state, jeopardizing its ability to secure a suitable overwinter site.

At a broader geographic scale, immatures of many species of wood-warblers (Parulidae) of eastern North America occur regularly in small numbers on the Pacific coast of California during fall migration (DeSante 1973). The regular occurrence of vagrant warblers along the west coast indicates that misoriented warblers have followed a direction that is a mirror image, across the north-south axis, of the correct direction (DeSante 1973). In a second part of DeSante's study, vagrant Blackpoll Warblers in orientation cages showed directed orientation behavior symmetric to both north-south and east-west axes. Such symmetry is compatible with the theory that vagrants display both the correct migratory direction and the mirror image of that direction. Apparently, immature warblers learn the stationary point of the night sky (celestial north) as a reference point to which to relate their inherited migratory direction (DeSante 1973). That vagrant migrants are typically immature birds is yet another indication that experience is likely to be important in the development of seasonally appropriate orientation in migratory birds (DeSante 1973).

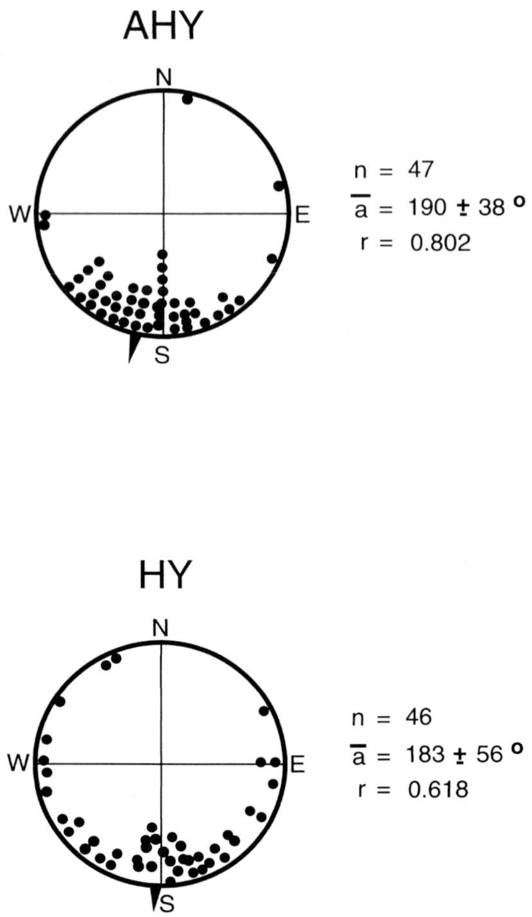

FIGURE 3. Mean orientation headings for experienced (AHY) and naive (HY) individual Savannah Sparrows. The mean headings (± angular deviations) and the r-values (measure of concentration of headings; a low value indicates dispersed headings, a high value indicates clustered headings) are given, and the mean headings are shown as arrows on the circumference of the circle. North is at the top (from Moore 1984).

SIGNIFICANCE OF AGE-DEPENDENT RESEARCH

Response to Energy Demand

The foregoing review of age-dependent research on migrant landbirds demonstrates that age-classes respond differently to the problems faced during migratory journeys. The *en route* ecology of migratory birds reflects a dynamic interaction between possible behavioral responses and a changing energetic state. For example, Red-eyed Vireos adjust their foraging behavior depending on their energetic status following trans-Gulf migration (Loria and Moore 1990). As a consequence of behavioral adjustments, lean birds are more likely to gain weight than fatter birds. Behavioral plasticity during the migratory period should come as little surprise given the different vegetation structures, wide variations in the resource quality and quantity, and changes in competitive pressures experienced during stopover (Morse 1971, Martin and Karr 1990). If flexibility in the foraging behavior of migratory birds provides a basis for adaptive responses to the energy demands of migration (Parrish *this volume*), a consideration of age as an independent variable becomes essential.

Experience and the Regulation of Migration

Conditions encountered *en route* and experience gained from previous migrations modify behavior and play a role in the regulation of bird migration (Gauthreaux 1978; Ketterson and Nolan 1985, 1988; Terrill 1988). For example, Gauthreaux (1978) and Ketterson and Nolan (1983) found differences in migration distance among age-sex classes of a wide variety of avian species. Ketterson and Nolan (1983) argue that several selective pressures influence such differential migration, including (1) higher mortality in young birds with no previous migratory experience, and (2) age-dependent trade-off between the greater mortality of longer migration and the increased survivorship associated with wintering in more tropical areas (i.e., the trade-off of adult migrants who are more experienced in satisfying energy demand is biased toward avoiding the risks of migration).

Experience, or lack thereof, has been implicated as a cause of age-specific differences in the stopover biology of migratory passerines (Morris et al. 1996, Woodrey and Moore 1997). Yet, distinguishing between lack of experience and constraint (sensu Curio 1983) can provide greater insight into the underlying mechanisms influencing age-specific migrant behavior. Whereas some aspects of migration may be endogenously "programmed" (Berthold 1984), young migrants might be constrained either through social dominance relationships or developmental forces (Marchetti and Price 1989). Observations of American Redstarts (Woodrey 1995) and Rufous Hummingbirds (Carpenter et al. 1993a,b) during stopover suggest that social constraints (i.e., density-dependent habitat use) are important in shaping migrant stopover behavior. However, regardless of the mechanism, age-related differences during the migration period can lead to differential recruitment into the breeding population.

Populations of long-distance passerine migrants appear to be limited during both summer and winter (reviewed by Sherry and Holmes 1993, 1995). For instance, long-term population data for American Redstarts suggests that breeding season events influence long-term population numbers (Sherry and Holmes 1992). In contrast, Baillie and Peach (1992) found that populations of several species of Palearctic-African migrants depend critically on events during the period between independence-from-parents to the birds' first breeding attempt.

Although the complex annual cycle of migrants makes it difficult to resolve when populations are limited (Morse 1980b, Holmes and Sherry 1988; Sherry and Holmes 1993, 1995), factors connected with migration and the stopover ecology of migrants should figure in any analysis of population limitation (Sherry and Holmes 1993, 1995). Three lines of evidence imply that migration events may limit populations: (1) the observation of territoriality in some species during migration (Rappole and Warner 1976; M. Woodrey, unpubl. data), (2) decreased insect abundance at stopover sites due to the presence of migrants (Moore and Yong 1991), and (3) occurrence of migrants among a variety of stopover habitats (Martin 1980, Moore and Kerlinger 1987, Moore et al. 1990). Further, migrants may be limited by food because of time constraints associated with locating suitable habitats during stopover (Moore and Simons 1992a, Moore et al. 1995). As Sherry and Holmes (1995:95) point out, "Such limitation could cause increased mortality of individuals in proportion to the size of migrants' populations, i.e., to density-dependent mortality during migration, although this phenomenon has yet to be demonstrated at this phase of the annual cycle." Future research in this area should address how different age-classes of migrants select and use stopover habitats, age-specific rates of predation pressure or mortality in habitats migrants select, and identification of intrinsic cues used by migrants to evaluate the suitability of habitats. These as-yet-to-be-collected data will be critical to our understanding of when during the annual cycle populations of birds are limited.

Conservation of Migratory Birds

Stopover biology is critical to the development of conservation strategies for migratory birds. Population declines of long-distance migrant passerines are linked with habitat loss on the wintering grounds (e.g., Wilcove and Terborgh 1984, Kelsey 1992, Sherry and Holmes 1995) and fragmentation of forested breeding habitats (e.g., Wilcove 1988, Bibby 1992; Sherry and Holmes 1992, 1995). The population status of migrants may also be linked to the availability of suitable *en route* habitat (Moore and Simons 1992a, Moore et al. 1993), where energy stores necessary for successful migration are available.

Habitat use during migration has profound consequences for a bird's (1) ability to satisfy the heightened energy demand of migration, (2) vulnerability to predators, and (3) exposure to environmental stress (Moore and Woodrey 1993). Unfortunately, little is known about habitat use by passerine migrants (but see Bairlein 1983, Moore and Kerlinger 1987, Watts and Mabey 1994, Moore and Woodrey 1993). As stopover habitats are converted or degraded, the cost of migration potentially increases and a successful migration is jeopardized. Protection of stopover habitats, used for feeding before and after they cross ecological barriers, is particularly important for the conservation of long-distance migrant passerines (Curry-Lindahl 1981). Because individual behavior influences the dynamics of populations, intensive study of the behavior and ecology of age-classes of intercontinental migrants during stopover is needed if we are to develop a successful conservation strategy for this group of birds.

ACKNOWLEDGMENTS

I thank F. R. Moore for the invitation to participate in this symposium. This manuscript benefited from the constructive comments of F. R. Moore, W. W. McDearman, S. T. Ross, J. T. Rotenberry, T. R. Simons, and members of the Migratory Bird Group: D. A. Aborn, J. B. Busby, D. A. Cimprich, J. Clark, M. P. Guilfoyle, and S. E. Mabey. Special thanks to T. W. Sherry and F. Spina for detailed reviews of an earlier draft of this manuscript and to L. D. Yates for her assistance with the preparation of this manuscript.

Studies in Avian Biology No. 20:53–70, 2000.

BEHAVIORAL, ENERGETIC, AND CONSERVATION IMPLICATIONS OF FORAGING PLASTICITY DURING MIGRATION

JEFFREY DAVID PARRISH

Abstract. Dietary plasticity is widespread and frequent in many landbird species during migration and typically involves shifts from stereotyped insectivorous diets during the breeding season to inclusion of other animal or plant matter, especially fruit. As with other forms of behavioral plasticity, flexibility in diet has probably evolved in response to environmental uncertainty, which, I argue, most landbirds encounter in terms of food resource availability during migration. The spatial and temporal uncertainty in insect availability during autumn stopover may have influenced the evolution of dietary flexibility during migration. Experiments and empirical observations from studies on Block Island, Rhode Island, off the northeastern coast of North America demonstrate that seasonal dietary shifts to fruit can strongly affect *en route* foraging behavior, habitat use, and migratory departure decisions. Migrants feeding on fruit use less expensive foraging behaviors, encounter more "prey" items per unit time, and perform fewer search movements than when feeding on insects. Furthermore, fruit removal experiments revealed that the presence of fruit influenced the habitats selected by frugivorous migrants such as Yellow-rumped Warblers (*Dendroica coronata*) during autumn. Because of seasonal frugivory, many species are selecting habitats that are different from those selected at other times of the year. Furthermore, dietary shifts also play a major role in migrant energy budgets during stopover by increasing energy intake while decreasing the energy expended during stopover foraging. Experiments with Red-eyed Vireos (*Vireo olivaceus*) and *Catharus* thrushes suggest dietary plasticity can increase energy intake and facilitate lipogenesis in species capable of assimilating novel diet types. Use of more temporally and spatially stable fruit resources may also decrease searching and handling time, and decrease energy expenditure during stopover foraging, resulting in positive net energy budgets for migrants. Aspects of migrant biology that change during the annual cycle, such as dietary shifts to fruit, must be integrated into conservation plans for landbird populations during migration.

Key Words: diet, energy budgets, foraging behavior, frugivory, plasticity, resource abundance, stopover ecology.

The behavioral ecology of nearctic-neotropical landbirds during migration is not a simple extension of the breeding or wintering season biology. Environmental and physiological pressures, such as increased energetic demand from nocturnal flights or unpredictable variation in habitat and resource availability, may be quite different during migration than at other periods during the annual cycle. These pressures have strong implications for the evolution of migrant behavior, life history, and population dynamics (Moore 1991a). One mechanism used by migrants to cope with these pressures is the ability to be flexible in behavior during migration. Migrating landbirds demonstrate this behavioral plasticity in a variety of ways, including flexibility in habitat selection, foraging maneuvers, and social interactions. An additional strategy used by many species is dietary plasticity, particularly visible as an expansion from insectivory during the breeding season to a diet including large proportions of fruit during migration. This dietary plasticity can profoundly influence the behavior, energy budgets, and ultimately the conservation of migrant landbirds during stopover periods.

BEHAVIORAL PLASTICITY

Behavioral plasticity has received extensive theoretical and empirical attention (e.g., Klopfer and MacArthur 1960, Klopfer 1967, Levins 1968, Stephens and Charnov 1982; Greenberg 1984a,c, 1987a, 1990; Ford et al. 1990, Martin and Karr 1990). Plasticity can be viewed as more than the simple antithesis of specialization (Morse 1980a), as has been suggested in the earlier use of the term (Klopfer and MacArthur 1960). Earlier, "plasticity" was used to define the resource or behavioral breadth of an organism, but Morse (1980a) first suggested that the concept of plasticity extends beyond the generalist and specialist dichotomy. He suggested that stereotypy and plasticity were the exploitation of resources under changing conditions in consistent and variable manners, respectively, and that plasticity was characterized by behavior that lacked long-term predictability in time or space. Greenberg (1990) proposed a further, functional definition: plasticity is the flexibility of organisms of the same genotype to vary in phenotype in the face of change. This definition of plasticity best suits the changing biology of nearctic-neotropical migrant landbirds, given their very diverse behavioral repertoire when experiencing spatio-temporally unpredictable environments.

Environmental instability can influence the evolution of behavioral and life history strategies through random shifts in direction and magnitude of selection pressures (Alerstam and Enckell 1979, Real 1980, Thompson 1991,

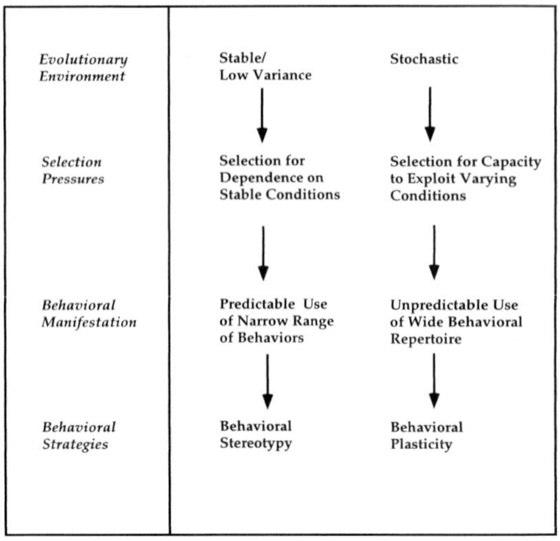

FIGURE 1. Evolutionary environment and selection pressures under which behavioral stereotypy or plasticity may evolve.

Schmitt 1994). Species faced with such frequent instabilities in selection pressure are left with few evolutionary behavioral options beyond the ability to diversify behaviors (Real 1980, Mangel and Clark 1988, Sherry 1990; Fig. 1). When confronted with uncertainty of expected fitness due to unpredictably changing environmental conditions, the optimal behavioral strategy for an organism may be the capacity to diversify the use of fitness-related behaviors—behavioral plasticity (Real 1980, Stephens and Charnov 1982, Real and Caraco 1986).

Environmental Uncertainty and Foraging Plasticity during Migration

Though the migratory journey itself may be obligate, migrating landbirds are confronted with great environmental uncertainty between the temperate breeding grounds and tropical wintering areas, and consequently exhibit behavioral plasticity during stopover periods (Martin and Karr 1990, Parrish 1997). After energetically demanding nocturnal flights, migrants are frequently forced to stop in unfamiliar habitats to refuel for subsequent migratory efforts. During this journey, habitats, weather conditions, and the quality and availability of some resources change unpredictably (Moore 1991a; Fig. 2). Although food resource availability at this time is typically declining (Fig. 2), it does so in a manner heavily dependent on environmental conditions at the precise time and location of foraging (Parrish 1996). Furthermore, extrinsic factors such as unpredictable weather (especially wind velocity and direction) may force migrants to fly undesirable trajectories and settle in unprofitable, or at least unexpected, stopover sites (e.g., Able 1977, Hutto 1985b, Lindström and Alerstam 1986, Moore and Simons 1992a). For example, the large abundances of migrants on coastal chenier islands of the Gulf of Mexico and glacial moraine islands off the New England coast are dependent on the frequent yet irregular fronts that displace migrating passerines and force birds to settle in areas that stray from "preferred" continental migratory paths (e.g., Able 1977, Moore et al. 1990, Morris et al. 1994, Parrish 1997). Moreover, resting and restoration of lipid reserves must occur under high, yet unpredictable predation risk. Several species of falcon appear to synchronize their migration with peak movements of migratory passerines, yet their abundance and distribution in any given location can not easily be predicted (Metcalfe and Furness 1984, Moore et al. 1990, Aborn 1994). The probabilities of surviving *en route* contingencies are diminished by the poor energetic condition of many passage migrants upon arrival at a stopover site. Low energy reserves and the need to replenish them limit the energy and time available for efficient habitat selection, resource acquisition, and predator avoidance during stopover (Moore et al. 1990). To cope with these constraints, migrants may exhibit behavioral plasticity during migration in their habitat selection (Parnell 1969, Bairlein 1983, Winker et al. 1992a), foraging behavior (Hutto 1981, Loria and Moore 1990, Martin and Karr 1990), and dietary shifts (Berthold 1976a,b; Wheelwright 1988, White and Stiles 1990, Willson 1991, Parrish 1997).

Frugivory throughout the Nonbreeding Period

Throughout the nonbreeding period many species of landbird migrants range widely in the degree of dietary plasticity. Records of diet shifts to frugivory, for example, are widespread throughout both the spring and autumn migrations and overwintering periods for a variety of taxa (Appendix). During autumn migration, some warbler species, such as American Redstarts (see Appendix for scientific names of all nearctic-neotropical migrants) at northern stopover sites, continue to feed when possible on insect types similar to those used during breeding (Parrish 1997). Other species may shift to use additional insect types, responding to diminishing nutrient demands of reproduction and the changing availability of insect prey during autumn, such as the shift from lepidopteran larvae to extensive use of Diptera and Homoptera by some Palearctic warblers (Bibby and Green 1981, 1983) or Hymenoptera by many neotropical migrants at a northern site (Parrish 1997).

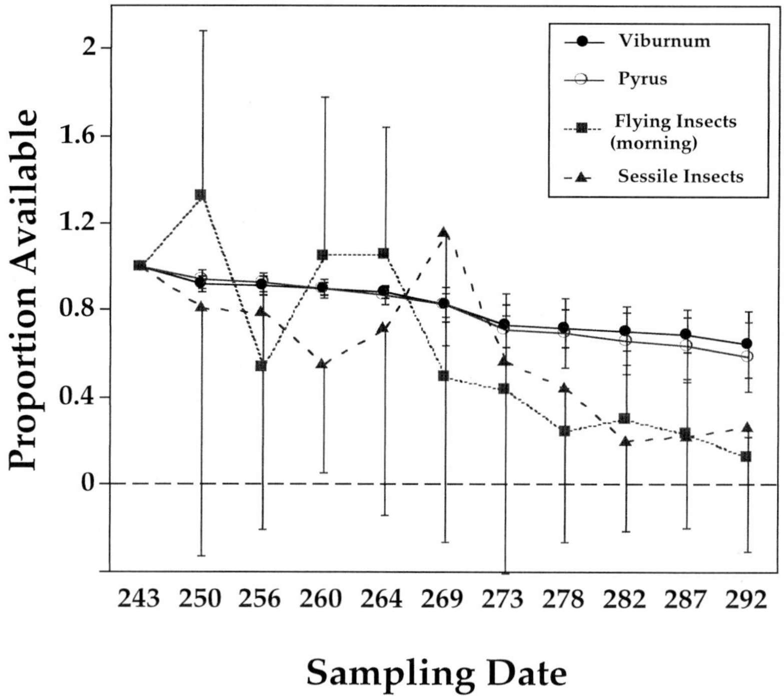

FIGURE 2. Decline in fruit and insect resources sampled at ten sites during autumn migration on Block Island in 1994. Fruit species sampled were *Viburnum recognitum* and *Pyrus melanocarpa,* and insects were sampled as flying insects (through sticky traps) and sessile insects (through branch fumigation counts). Error bars represent one standard deviation and demonstrate high variances about the mean for insect resources throughout the season. Sampling dates are 31 August (Julian date 243) through 19 October (Julian date 292).

Still other migrants may begin to feed on non-insect arthropods, e.g., marine amphipods and other invertebrates taken by Yellow-rumped Warblers late in migration (J. Parrish, pers. obs.). Perhaps the most widespread pattern of diet shifts, however, is that from largely insectivorous diets to include high proportions of fruit, as is common among many migrant thrushes (Turdidae), vireos (Vireonidae), mimids (Mimidae), and wood-warblers (Parulidae). Such extensive frugivory during autumn migration has been widely documented in both palearctic and nearctic migration systems (e.g., Izhaki and Safriel 1985, Johnson et al. 1985, Jordano 1988, White and Stiles 1990, Levey and Stiles 1992, Parrish 1997). Because diet shifts to fruit represent an inclusion of resources differing extensively in distribution and nutritional content (Table 1), and because these shifts are almost complete in some migrating landbird species, they present an important opportunity to explore the consequences of seasonal dietary changes for the be-

TABLE 1. Qualitative Comparison of Relative Nutrients and Distribution of Fruit and Insect Resources for Migratory Landbirds (Summarized from Moermond and Denslow 1985)

Character	Fruit	Insects
Protein content	Low	High
Carbohydrate content	High	Variable
Fat content	Low in northern latitudes (some species with high lipid content)	High
Rate of autumnal decline	Slow, present through much of autumn	Rapid, variable and unpredictable
Distribution	Clumped; patchy	Variable; not clumped. High spatial and temporal variance
Ease of capture	High	Low
Detectability	High	Variable

TABLE 2. DIET OF RED-EYED VIREO AND *CATHARUS* THRUSHES DURING AUTUMN MIGRATION ON BLOCK ISLAND, RHODE ISLAND, ON THE NORTHWESTERN ATLANTIC COAST (1993–1995), AND DURING SPRING MIGRATION ON HORN ISLAND, MISSISSIPPI, ON THE GULF OF MEXICO COAST (1994–1995) BASED ON ANALYSES OF FECAL SAMPLES FROM MIST-NETTED BIRDS

	Autumn migration				Spring migration		
Species	No. fecal samples (N)	No. samples with insects (%)	No. samples with fruit (%)	Mean % fruit per sample	No. fecal samples (N)	No. samples with insects (%)	No. samples with fruit (%)
Red-eyed Vireo	194	185 (95)	194 (100)	73.9	311	246 (79)	158 (51)
Veery	15	9 (60)	15 (100)	82.7	98	67 (68)	70 (71)
Gray-cheeked Thrush	8	5 (63)	8 (100)	85.0	41	30 (73)	32 (78)
Swainson's Thrush	23	16 (70)	22 (96)	80.0	103	65 (63)	84 (82)
Hermit Thrush	94	67 (71)	94 (100)	83.6	9	4 (44)	7 (78)
Total	334	282 (84)	333 (99.7)		562	412 (73)	351 (62)

Notes: Mean percent fruit per sample during autumn migration based on visual approximation with microscope to nearest 10%. Spring migration data from J. Clark and F. Moore (unpubl. data).

havior, energetics, and conservation of nearctic—neotropical migratory landbirds.

Frugivory is not limited to the temperate zone during autumn migration. Some neotropical migrant species remain frugivorous along tropical migratory routes during autumn even as insects become more abundant at more southerly latitudes during the tropical wet season. For example, during their southern migration throughout southern Mexico, Red-eyed Vireos consume fruits of *Phytolacca rivinoides,* a tropical shrub whose fruiting phenology appears timed with nearctic breeding landbird migrations (Winker 1995). Moreover, the omnivorous diets of overwintering neotropical migrants within the tropics are well documented (e.g., Howe 1977; Greenberg 1981, 1993; Wheelwright et al. 1984; Blake and Loiselle 1991, 1992a; Poulin et al. 1994; see Appendix). However, most investigations in the Neotropics of migrant frugivory have been limited to the roles of overwintering landbirds as seed dispersers (e.g., Olson and Blum 1968, Leck 1972, Howe 1977, Howe and DeSteven 1979) or the influence of fruit on wintering habitat use (Willson et al. 1982, Martin 1985, Blake and Loiselle 1992b). Information is now available for a few species of the importance of fruit in meeting the energy demands of transient migrants within the tropics (Morton 1973, Greenberg et al. 1995a).

During the northward spring migration in the tropics, some neotropical migrant species such as Wood Thrushes and Swainson's Thrushes are reported to shift diets to fruits that become abundant during the tropical dry season, presumably to satisfy the need for lipogenesis (Martin 1985, Blake and Loiselle 1992a). Spring frugivory can also occur within the temperate zone, but this phenomenon and its importance are understudied (Appendix). Where fruits are available as a result of late winter fruiting plants or remnant fruits from the preceding autumn, many migrants may continue to utilize fruits in attempts to meet energetic demands during spring migration (Willson 1991; J. Clark and F. Moore, unpubl. data; Table 2, Appendix). With selective pressures for early arrival on the breeding grounds for acquisition of mates and territories (Francis and Cooke 1986, Marra et al. 1998), earlier spring migration prior to adequate or reliable insect emergence may be possible through dietary supplementation with spring fruits.

Data from the coast of the Gulf of Mexico suggests that spring frugivory is quite different from that of autumn migration (J. Clark and F. Moore, unpubl. data). For example, fruits are not as abundant among patches in spring, and are generally poorer in nutritional quality compared to autumn (J. Clark, F. Moore, and J. Parrish, unpubl. data). Perhaps as a result of these resource differences and the time constraints of spring migration (Francis and Cooke 1986, Morris et al. 1994, Marra et al. 1998), neotropical migrants do not appear to gain significant mass during spring frugivory compared to autumn (Parrish 1997; J. Clark and F. Moore, unpubl. data). Moreover, 22 of 25 species demonstrating spring frugivory at this southern temperate latitude showed a pattern of more extensive frugivory in fat migrants than in lean migrants (J. Clark and F. Moore, unpubl. data).

Although fruit may be available to migrant landbirds at some sites in spring, many species during the northward migration appear to be returning to insectivorous diets typical of the breeding season, thereby showing the mirror image of autumnal diet shifts (Table 2, Appendix). Yet, in some species (e.g., Northern Waterthrushes, Yellow Warblers, and Common Yellowthroats) frugivory appears common during migration and less frequent during the wintering period, suggesting an advantage to seasonal dietary plasticity during migration (Parrish 1997; Appendix). That in many species there exists some level of frugivory during most of the annual cycle suggest that neotropical migrants may be more correctly viewed as omnivores that exhibit only seasonal stereotypy on insects during reproduction, while being plastic in their diets during the remainder of the year (Levey and Stiles 1992, Levey 1994). This view has previously been suggested by Levey and Stiles (1992), who proposed that nearctic-neotropical migrant landbirds are descended from neotropical taxa that exhibit seasonal altitudinal or intratropical migration in response to changing fruit and nectar availability at forest edge, canopy, and seasonally dry habitats. The tracking of seasonal fruit and nectar resources, they suggest, may have been influential in the very evolution of the nearctic-neotropical migration system of passerine birds (Levey and Stiles 1992). This broader view of neotropical migrants as tropical omnivores in both their evolutionary origin and in their behavioral ecology during the majority of the annual cycle (Levey 1994), may be a more appropriate context in which to explore the implications of dietary plasticity during migration.

Constraints on Dietary Plasticity

That migrating species vary in extent of dietary plasticity shown during the nonbreeding period suggests that certain physiological, morphological, or environmental constraints determine the capacity to exploit novel diet types. First, nutritional requirements of migrant species change over the annual cycle. For example, needs for dietary protein, particular amino acids, and minerals probably change between breeding (for the production of eggs and feeding of young) and migration (Izhaki 1992). However, most migrant populations experience some depletion of muscle mass during migratory efforts (Lindström 1991, Lindström and Piersma 1993), as well as partial or full replacement of plumage prior to or during migration (King 1974, Winker et al. 1992a). Protein needs during migration may therefore still exist, albeit to a lesser degree. Moreover, diet choice for the combination of nutrients and secondary compounds (Izhaki and Safriel 1989, Izhaki 1992) that optimizes lipogenesis in migrating birds should be under heavy selective pressure during stopover. The morphology and physiology of a taxon can further limit its capacity to exploit new diet types, thereby determining the extent of dietary plasticity. For example, certain birds may be unable to process seeds or certain fruit sizes because they lack the bill dimensions or jaw musculature for access to and ingestion of available resources (Moermond and Denslow 1985, Jordano 1987), or because gut length (Jordano 1987) or digestive strategies (Martinez del Río and Karasov 1990, Levey and Duke 1992, Place and Stiles 1992, Levey and Karasov 1994) control the assimilation of ingested foods. Moreover, these physiological and morphological characters have a phylogenetic context which can limit the capacity of certain taxa for dietary plasticity.

Psychological factors and social interactions may also play a role in determining the extent of dietary plasticity. Certain warbler and sparrow species, for example, exhibit an intrinsic "neophobia," or an aversion to novel environments or objects (see Greenberg 1990 for review). Such psychological bases for the limitation or expansion of foraging behavior appear qualitatively related to the extent of frugivory shown during the nonbreeding season: the neophobic Chestnut-sided Warbler seems less likely to demonstrate frugivory in the tropics than its neophilic congener, the Bay-breasted Warbler (Greenberg 1979). Similar patterns appear to exist in other *Dendroica.* For example, Black-throated Blue warblers are highly plastic in their diets during migration and the overwintering period, and they appear more flexible in foraging site selection during breeding periods than congeneric Black-throated Green Warblers, which remain predominantly insectivorous during migration and are more stereotypic in microhabitat selection in both the tropics and the breeding season (Whelan 1989, Greenberg 1992, Parrish 1995b; C. J. Whelan and J. D. Parrish, unpubl. data). Competition for resources among and between species can further limit the possibilities of using additional resource types during stopover (Sealy 1988, 1989; Moore and Yong 1991).

The nutritional composition and distribution of resources during migration, however, may be most decisive in determining which species are plastic, why this plasticity has evolved, and what consequences dietary plasticity may have for the conservation of migratory landbirds during migration. Fruit and insect resources differ in their nutrient composition and distribution in space and time during autumn, the principal migration season characterized by dietary shifts to frugivory in the northern temperate zone (Table 1; but see discussion of spring frugivory above). Insects, in contrast to most fruits, are generally high in protein, but vary extensively in the amounts of carbohydrate per gram relative to fruits (Morton 1973, Moermond and Denslow 1985). Moreover, most fruits available to *en route* landbirds in northern temperate stopover areas are lower in percentage of digestible lipids than available insects (Stiles 1980a, Johnson et al. 1985, Moermond and Denslow 1985, Borowicz 1988; but see Conway et al. 1994). Thus fruits may present a valuable source of carbohydrates, but a variable reward in lipid composition between plant species (Morton 1973, Johnson et al. 1985). These fruit species can also vary in nutrient, caloric, and water content over the season (Stiles 1980a, Johnson et al. 1985, White 1989). The distribution of fruit and insect resources are also extremely different in ways that influence the relative value of the two resource types for migrants. Fruits are typically patchily distributed and spatially aggregated over a variety of scales. Individual infructescences in some plants such as *Viburnum* spp. and *Sambucus* spp. can hold over 1000 individual fruits, and these plants also tend to occur in aggregated groups within a habitat (Levey 1988, 1990), perhaps as a result of avian dispersal effects on seed shadows (Levey 1991, Willson and Whelan 1990). Insects, in contrast, are more variable and widely distributed than fruits (Wolda 1978, 1988), especially during autumn in northern temperate zones (Fig. 2; J. Parrish, unpubl. data). Insects are also more difficult to detect due to cryptic coloration or mimicry, as opposed to the brightly colored displays of fruiting shrubs that serve as advertisement for seed dispersal (Willson and Thompson 1982, Willson and Whelan 1990, Willson et al. 1990). Furthermore, insects are usually active and mobile, and often possess escape mechanisms to evade predation, whereas fruits are sessile and present little difficulty for a hyperphagic migrating landbird. Temporal distributions of insects and fruit also vary over the autumn period, with the decline in insect resources being much more rapid and unpredictable over the migration period than that of fruit resources (Fig. 2; J. Parrish, unpubl. data). I suggest that these differences in display and spatial distribution render insects a much more "expensive" foraging resource than fruits during autumn migration in terms of searching time, handling time and energy expenditure (Parrish 1996).

Debate Over the Importance of Fruit to Migrating Landbirds

Despite the extensive frugivory noted in many migrating species, many researchers have suggested that fruit plays only a minor role in lipid deposition by migrant birds during stopover periods (Berthold 1976a,b; Izhaki and Safriel 1990). In early feeding experiments, Berthold (1976a,b) found that any mass gain by captive Blackcaps (*Sylvia atricapilla*) during the experimental period was closely tied to a decrease in ingested vegetable matter and an increase in the proportion of animal food in the diet. Moreover, when European Blackbirds (*Turdus merula*), European Robins (*Erithacus rubecula*), and Garden Warblers (*Sylvia borin*) were fed exclusively fruit diets, they suffered precipitous declines in body mass and fat levels that were only reversed when they were returned to animal diets. Berthold concluded that his findings were incompatible with the hypothesis that fruits are important to migrant passerines during migration, allowing only that fruit may be beneficial as a supplementary food for transient birds lacking any other food.

There is conflicting evidence, however, which suggests that fruit is of greater importance to the energetic condition of migrating songbirds than proposed by earlier workers. The occurrence of frugivory in *en route* migrants is frequent and extensive in both the Palearctic (Mead 1966, Blondel 1969, Fry et al. 1970, Ferns 1975, Thomas 1979; Herrera 1981, 1984; Jordano 1981, Stoate and Moreby 1995) and the Nearctic (Baird 1980, Stiles 1980a, Rybcyzynski and Riker 1981, Johnson et al. 1985, Parrish 1997). More importantly, there are potential associations between avian use of fruit and fruiting habitats, and higher body mass. For example, Thomas (1979) showed that frugivorous Garden Warblers feeding on the fig *Ficus carica* in southern Portugal during migration weighed significantly more than insectivorous conspecifics feeding in reed beds at the same site. Yet most convincing is the evidence presented by Simons and Bairlein (1990) and Bairlein (1990), who experimentally investigated the contradiction between Berthold's findings and the high incidence of frugivory in palearctic migrants, using Garden Warblers feeding on lipid-rich fruits. Birds were able to gain mass on fruit diets, suggesting that non-animal food resources may be more important to stopover mass gain than was once considered. Indeed, Willson (1991) has called for a renewal of interest and a re-questioning of the importance of fruit to the biology of passage migrant landbirds.

BEHAVIORAL AND ENERGETIC CONSEQUENCES OF FRUGIVORY DURING MIGRATION—A CASE STUDY AND EXAMPLES

Flexible, plastic diets that include frugivory can have direct consequences for nearctic-neotropical landbird migrant foraging behavior, habitat selection, and energy budgets during *en route* stopover. The influence of diet on these facets of migrant biology has been the focus of empirical observation studies and experimental work during autumn migration on Block Island, Rhode Island.

Study Site and Methods

Block Island is a 2900 ha glacially deposited landmass 18 km off the coast of southern New England, where northwesterly winds associated with cold fronts force migrants to concentrate on offshore islands (Able

1977). The site is composed of three basic habitat types: contiguous northern bayberry (*Myrica pensylvanica*) from 1 to 2.5 m in height, maritime scrub (from 2–4 m high and predominantly *Pyrus melanocarpa, Viburnum recognitum, Amelanchier* spp., *Myrica pensylvanica, Rosa* spp., and *Rubus* spp.), and older orchard scrub (measuring 4–7 m and composed of maritime scrub species, but also with high densities of *Prunus serotina* and *Pyrus malus*). Ten 12-m mist nets (32-mm mesh) were used in the maritime scrub habitat to sample individual migrants for diet, experimentation, and general monitoring of stopover population ecology. I focused these studies of seasonal frugivory on the maritime scrub habitat because of its high density yet relatively low diversity of fruit and fruiting shrubs, which has facilitated experimental approaches toward understanding diet shifts.

Red-eyed Vireos and *Catharus* thrushes (*C. ustulatus, C. minimus, C. fuscescens,* and *C. guttata*) were chosen as focal species for experiments because of their abundance, ease of acclimation to captivity, and extremely frugivorous diets. I determined the trends in body mass and fat condition of these focal species during autumn migration by analyzing capture data from the same location. Analyses of energetic condition changes were made on recaptured individuals netted on Block Island from 1969–1995 by Mrs. F. D. Lapham and the author (Table 3). I conducted diet surveys of the focal species with individuals netted during the autumns of 1993–1995. Birds were removed from mist nets and placed into transport bags (Parrish et al. 1994), which allowed collection of fecal samples of captured birds to estimate later the degree of frugivory to the nearest 10% in bird diets before capture (e.g., Jordano and Herrera 1981). I acquired large numbers of samples with this technique, providing an estimate of frugivory for focal experimental species (Table 2). The patterns of dietary data (Table 2) and energetic condition (Table 3) for these focal species suggest that they were gaining mass in the field on highly frugivorous diets while resting on Block Island.

Because of potential, yet undocumented, differences in the temporal distribution of fruit and insect resources that could influence the behavior and energetics of *en route* migrants, I documented the relative decline over the autumn migration season of the two resources on Block Island during 1994. Flying insects in the 2–4 m coastal scrub habitat were monitored during morning and afternoon intervals with Tanglefoot®-coated boards (22 cm × 22 cm; Cooper and Whitmore 1990, Wolda 1990, Kuenzi et al. 1991) placed at 1-, 1.5-, and 2-m height intervals at ten points spread through the study area. I also monitored sessile insect abundance by bagging randomly chosen, 0.5-m branches from northern bayberry and northern arrowwood (*Viburnum recognitum*) at the same ten points and fumigating them with permethrin (Morse 1976, Wolda 1990). Fruit crops (ripe fruits per designated branch segment) of tagged 0.5-m branches of northern arrowwood and black chokeberry (*Pyrus melanocarpa*) were also monitored. Fruit abundances on five branches of each plant species were followed throughout the season at five of the insect monitoring points in the study site. Resource measurements were made

TABLE 3. ENERGETIC CONDITION CHANGES AND STOPOVER LENGTH OF *Vireo Olivaceus* AND *Catharus* THRUSHES ON BLOCK ISLAND, RHODE ISLAND, DURING AUTUMN MIGRATION BASED ON INDIVIDUALS RECAPTURED WITHIN ONE SEASON BETWEEN 1970–1995

Species	N	Mass (g) first capture	Mass (g) last capture	t	Fat score first capture	Fat score last capture	t	Stopover length (days)
Red-eyed Vireo	308	18.3 ± 2.5	19.9 ± 3.3	−10.9***	1.5 ± 2.5	1.9 ± 1.1	−9.0***	4.5 ± 4.4
Veery	95	32.9 ± 3.7	35.7 ± 5.1	−6.3***	1.2 ± 1.0	1.9 ± 1.1	−6.6***	4.3 ± 3.0
Gray-cheeked Thrush	98	33.5 ± 4.3	37.3 ± 5.9	−8.2***	1.6 ± 1.2	2.4 ± 1.2	−8.0***	4.3 ± 28
Swainson's Thrush	110	30.6 ± 3.1	33.2 ± 4.3	−6.7***	1.0 ± 0.9	1.9 ± 1.2	−6.7***	5.1 ± 4.0
Hermit Thrush	121	30.0 ± 2.4	31.2 ± 5.8	−2.6*	0.8 ± 0.8	1.3 ± 1.0	−6.9***	4.6 ± 3.8

* = $P < 0.05$; *** $P < 0.001$.

on average every four days during the period from 2 September–25 October.

RESOURCE VARIATION

Resource abundance data from 1994 (Fig. 2) indicated dramatic differences in the overall rate of decline of insect and fruit resources available to migrants at this northern temperate stopover site. When expressed as a percentage of the original count remaining during the season, insects declined at a much faster rate over autumn than did fruits. Moreover, insect resources were more stochastic in their temporal availability during the season, an expected pattern given the dependence of many orders on appropriate temperature and wind regimes for flight (Wolda 1988). The variance among ten sampling sites, represented in Fig. 2 by standard deviation error bars at each temporal point, illustrates the high spatial variability of insect resources relative to the two principal fruit species used by most landbird migrants. Thus for a fat-depleted, inexperienced, recently arrived migrant, fruit resources may be more reliable in both space and time than are insects.

BEHAVIORAL IMPLICATIONS

Shifts in diet to fruit can influence many facets of migrating songbird behavior in ways that affect energetic and habitat requirements as well as migratory strategies during migration. For example, the foraging behavior of many landbird species during stopover is characterized by the use of an expanded range of foraging maneuvers (Martin and Karr 1990). Dietary plasticity can produce these observed expansions in foraging behavior, whether they are increased ranges of substrates from which prey are taken, an expanded repertoire of foraging maneuvers, altered foraging rates, or shifted foraging-site selection. When diet is altered, foraging behavior may subsequently change in response to the novel distribution and accessibility of new prey items, thereby explaining observed differences in foraging activity during stopover periods. For example, the clumped and stationary distribution of fruits may lower foraging rates and increase stationary foraging by frugivorous migrants, while allowing them to use energetically less expensive non-aerial maneuvers with greater frequency than is required during insectivory. To test the hypothesis that foraging behavior would differ between fruit and insect diets during stopover, foraging *en route* migrants were observed on Block Island during 1994. Search and attack rates and proportions of different foraging behaviors used (based on methods of Remsen and Robinson [1990]) were determined for each diet type from 5 September through 16 October as part of a larger study of migrant foraging behavior (Parrish 1996). Birds were classified as foraging on fruit or insects based on observations of the resources at which foraging maneuvers were directed; such classification was possible since birds rarely switched between insect and fruit resources within one foraging bout observation (Parrish 1996). Based on 372 observed foraging sequences, search rates were lower and attack rates were higher for fruit diets than for insect resources, suggesting more frequent "prey" encounters as a result of the clumped nature of fruit (search rates: $t = 4.47$, $P < 0.001$; attack rates: $t = 4.90$, $P < 0.001$; Fig. 3). Moreover, energetically costly aerial attack maneuvers were used more frequently during insectivory than during frugivory ($\chi^2 = 31.45$, $P < 0.001$; Fig. 3). These data suggest that diet shifts to fruit may afford landbirds that are physiologically capable of dietary plasticity the opportunity to maintain caloric intake with lower energetic expenditure during foraging. Thus diet can affect foraging behavior in ways that may, in

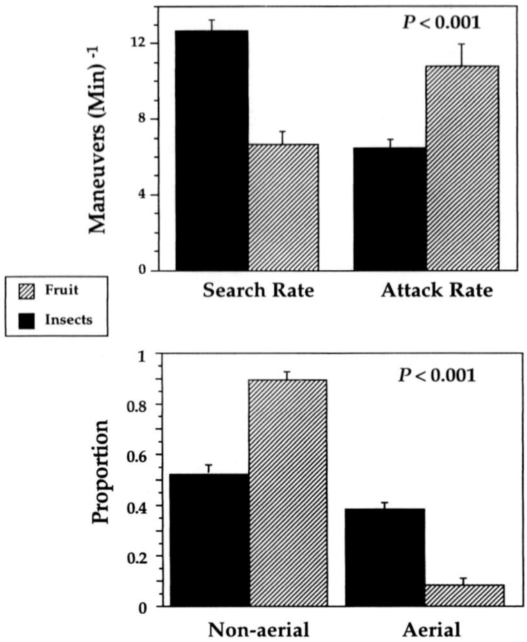

FIGURE 3. Foraging behavior of migrant landbirds on fruit and insect diets during autumn, 1994, on Block Island, Rhode Island. Foraging behaviors observed include the average search and attack rates of foraging birds, and the proportion of aerial versus non-aerial attack maneuvers used by migrants on fruit and insect diets. Aerial maneuvers included sallies, sally-hovers, sally-strikes, and leaps, whereas non-aerial maneuvers consisted of gleans, reaches, hangs, and lunges, according to Remsen and Robinson (1990) and Parrish (1996).

turn, influence energy balance during stopover periods.

With shifts in diet choice during the nonbreeding season, many species may also shift their habitat selection behavior. For example, work from both the tropical wintering grounds and temperate migration sites suggests that frugivorous migrants may select nonbreeding habitats on the basis of fruit availability alone. In the Neotropics, Blake and Loiselle (1992b) found that numbers of frugivorous overwintering migrants in Costa Rican rainforest were greatest in the sites where fruit was most prevalent. Further work by Martin (1985) revealed that migrating birds in Panama concentrated in second-growth forests perhaps due to the greater abundances of accessible fruits in those areas. In the Palearctic, wintering European warblers were also found in greater densities in habitats with the greatest abundance of fruits, to the extent that Herrera (1985) suggested that those migrants, as seed dispersers, were significant modifiers of their own habitats. Large-scale experiments with wintering distribution of palearctic migrants in olive orchard landscapes of southern Spain have demonstrated that overwintering landbirds track fruit availability and that migrant and fruit abundances are closely linked (Rey 1995).

Less well-studied is the relationship between novel diet types, such as fruit, and the habitat selection of *en route* migrants. Bairlein (1983) hypothesized that the change in habitats used by Blackcaps on migration in Germany resulted from including more fruit in their diet during migration than during the breeding season. Bibby and Green (1981, 1983) showed that a more specific habitat use of marshland reed beds was correlated with a diet shift in Sedge Warblers (*Acrocephalus schoenobaenus*) to the plum-reed aphid *Hyalopterus pruni* during migration. Moreover, Garden Warblers on migration occurred in highest densities in the presence of *Ficus carica* in southern Portugal (Thomas 1979).

Yet the influence of fruit on stopover habitat selection in nearctic-neotropical migration systems has not been as thoroughly explored. Blake and Loiselle (1991) noted a shift in the abundance of temperate migrants in tropical lowland habitats of Costa Rica simultaneous with the peak of fruit abundances. In a comparative study between northern temperate zones during migration and Panamanian rainforest, Willson et al. (1982) showed migrating frugivores at the northern migration site to be more common in light gaps, where fruits were more abundant than in the forest interior. In other correlative studies, Martin and Karr (1986) and Blake and Hoppes (1986) showed that habitat selection of frugivorous *en route* migrants was correlated with forest gap understory sites, where fruit abundance was concentrated during autumn. I have found similar patterns for certain species during autumn stopover in coastal maritime scrub habitats on Block Island, where fruiting shrub densities are exceedingly high (J. Parrish, unpubl. data). For example, Red-eyed Vireos, a highly frugivorous species (Table 2), are captured over ten times more frequently in coastal maritime scrub than in old orchard habitat on Block Island, with the former habitat type having a greater density of fruiting shrubs (t-test, $P < 0.05$). Furthermore, the Yellow-rumped Warbler, a species that feeds heavily on northern bayberry fruits during the winter, also appears to base its habitat use on fruit resources. This migrant species possesses specific adaptations for processing of the fatty esters surrounding bayberry pericarp (Yarbrough and Johnston 1965, Wilz and Giampa 1978, Place and Stiles 1992). However, these warblers actually begin their diet shift while on migration through northern temperate latitudes (Parrish 1997), and occur in greatest densities in habitats where bayberry plants predominate (J. Parrish, unpubl. data). On Block Island, overall migrant capture rates were higher in bayberry habitats than in coastal shrubland during autumn migration, 1995 (Wilcoxon sign rank test, $z = -2.67$, $P = 0.008$; Fig. 4). This pattern was most apparent in the latter half of the migration period (after the first week in October at the site; calendar date = 282, Fig. 4), when Yellow-rumped Warbler abundances increased rapidly and typically constituted over 35% of all daily migrant captures on Block Island.

Moving beyond correlational analyses, I have conducted fruit removal experiments in coastal shrubland habitats to determine if the presence of fruit in conjunction with seasonal frugivory is responsible for habitat use during migration. I removed all bird-dispersed fruits from 30 m × 30 m plots and mist netted migrants within control and removal plots to compare resulting bird abundances. Results indicated that migrant habitat use by highly frugivorous species was strongly influenced by the presence of fruit in a habitat. For example, Yellow-rumped Warblers (which are highly frugivorous) were more abundant in control plots within bayberry habitats, where fruit was present, than in treatment plots, where fruit had been removed (Wilcoxon sign rank test, $z = -3.008$, $P = 0.003$; Fig. 4). Few such experiments have been conducted, yet they are critical to determine the habitat selection mechanisms used during migration and the extent to which these are modified by dietary plasticity (Morse 1985).

Migratory strategies of *en route* migrants, including departure decisions and, therefore, stop-

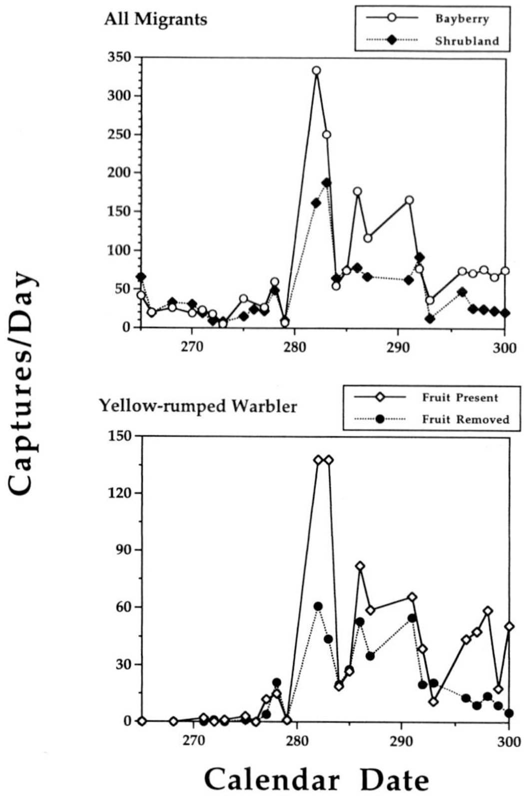

FIGURE 4. Habitat use by migrant landbirds during autumn, 1995, on Block Island, Rhode Island. Top figure represents the total number of captures of all migrant species in bayberry and coastal shrubland species. Bottom figure illustrates responses (distribution of mist-net captures per day) in habitat use by highly frugivorous Yellow-rumped Warblers to a fruit removal experiment within bayberry habitats.

over length, can also fluctuate with seasonally changing dietary needs. Such an effect is proximately controlled by time and energetic constraints, which ultimately can be influenced by the food resources used during migration. For example, Palearctic *Acrocephalus* warblers show differences in dietary plasticity during migration (Bibby and Green 1981, 1983). Sedge Warblers are stereotyped in their insectivorous diets, feeding on plum-reed aphids. As a result of this narrow, stereotyped diet, Sedge Warblers migrate more rapidly, gain more mass at each stopover site, and fly further per migratory flight than the more dietarily plastic Reed Warbler (*A. scirpaceus*; Bibby and Green 1981). Bibby and Green suggested that the degree of dietary plasticity during migration was influential in the evolution of migration strategies by these congeners.

In general, migrant stopover length is inversely proportional to energetic condition on arrival; that is, migrants will continue their journey (exhibit migratory flight behavior) once they have gained sufficient mass (Biebach et al. 1986, Moore and Kerlinger 1987, Loria and Moore 1990, Lavee et al. 1991, Kuenzi et al. 1991, Yong and Moore 1993). Others (Rappole and Warner 1976, Gwinner et al. 1984, Terrill 1990) have suggested that birds that are unable to replenish fat reserves in a stopover location will depart the site sooner in search of better foraging opportunities. Dietary plasticity can facilitate or slow migrant weight gain during stopover, depending on individuals' abilities to process the novel diet types and the availability of seasonally preferred foods. I therefore predicted that diet could proximately influence the migratory strategies of frugivorous landbirds as measured by the caged migratory restlessness activity (*Zugunruhe*) of focal individuals. I measured the nocturnal activity of experimental Red-eyed Vireos (N = 80 birds) and *Catharus* thrushes (N = 59 birds) between sunset and sunrise in holding cages with electronic activity-recording perches. Birds were subjected to four-day ad libitum diet treatments of exclusively insects (mealworms, *Tenebrio molitor*), exclusively fruit diets (fruits of *Phytolacca americana, Viburnum recognitum,* and *Pyrus melanocarpa*), and a mixed control diet of all food items. Red-eyed Vireos showed a strong inverse relationship between energetic condition and migratory behavior by increasing their nocturnal activity significantly when placed on fruit diets, which lowered their energetic condition (activity experiments analyzed by ANOVA; treatment effect: $F = 5.88$, $P < 0.01$; Figs. 5–6). This response suggests departure behavior in search of more suitable sites where mass gain is more assured (sensu Terrill and Ohmart 1984, Terrill 1990). In contrast, *Catharus* thrushes showed migratory activity independent of diet types (ANOVA; treatment effect: $F = 0.16$, $P = 0.85$; Fig. 5). These results demonstrate the influence of dietary plasticity on the migratory strategy of the Red-eyed Vireo, but suggest that fruit diets are essentially equal to insect diets with regard to factors influencing *Catharus* thrush migratory strategies.

Energetic Implications

Diet shifts occurring during migration can have profound implications for energy budgets of passage landbirds by influencing rates of mass change, optimal fat load, and energetic condition at departure. Although most studies of stopover ecology are concerned with the importance of energy *intake* during stopover in terms of hyperphagia (Loria and Moore 1990), optimal *en route* foraging (Moore and Simm 1985), and alternative diet choice (e.g., Berthold 1976a,b;

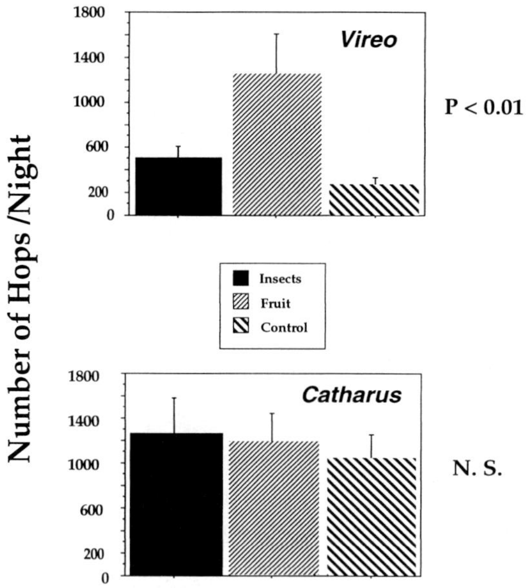

FIGURE 5. Mean (± SD) nocturnal migratory activity in Red-eyed Vireos and *Catharus* thrushes fed four-day, ad libitum insect diets (*Tenebrio* larvae), fruit diets (*Viburnum recognitum, Phytolacca americana,* and *Pyrus melanocarpa*), or a control diet of fruits and insects. Activity was measured as the mean number of perch hops recorded in caged birds between sunset and sunrise, and was analyzed using analysis of variance.

Graber and Graber 1983; Bairlein 1990), few have considered the additional factor involved in any net energy budget equation: energy *expenditure* during stopover. Graber and Graber (1983) suggested some of their spring foraging observations of warblers in areas with insufficient food resources were consistent with the concept of energy conservation during *en route* foraging. Additionally, for migrant hummingbirds, which have different physiological options for energy budget maintenance, Hiebert (1991) has suggested the importance of maximizing energy conservation and minimizing time for premigratory fattening through seasonal use of torpor. Yet in migratory passerines, little consideration has been given to the role of minimizing energy expenditure in realizing net positive energy budgets during stopover. Shifts to diets that help maximize energy intake and minimize energy used during stopover would thus be strongly favored under selection for optimal behavioral strategies in *en route* migrants.

Despite the controversy over the abilities of migrants to maintain themselves on fruit (Berthold 1976a,b; Izhaki and Safriel 1989, Mack 1990, Simons and Bairlein 1990, Izhaki 1992), neotropical migrants maintain high levels of frugivory during stopover and gain mass in the field (e.g., Johnson et al. 1985; Tables 2, 3). Most fruits are generally low in protein and lipids but high in carbohydrates (Snow 1971, Morton 1973, Moermond and Denslow 1985, Debussche et al. 1987, Herrera 1987). However, Bairlein (1985b, 1987a) has shown that captive Garden Warblers recover from an initial mass loss on forced low-protein diets (simulating frugivory) by increasing their daily food intake and improving assimilation efficiency of fat and carbohydrates. Moreover, birds fed high carbohydrate diets when under fat-reduced diet treatments did not lose body mass. Subsequently, Bairlein (1990) also demonstrated experimentally that laboratory Garden Warblers could gain mass on fruit diets during migration through a series of changes in fruit selection and physio-

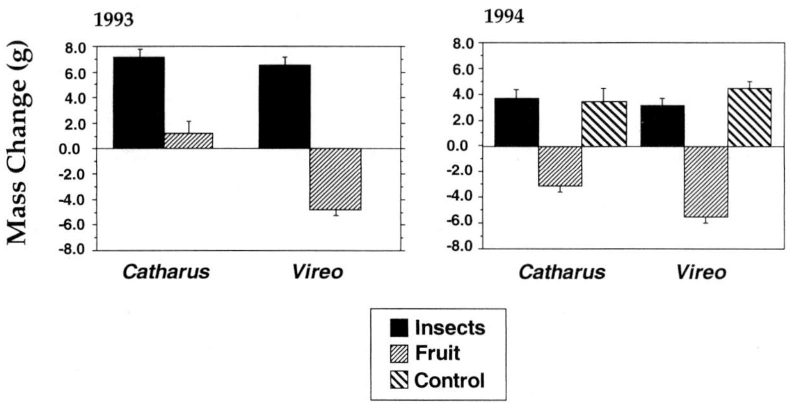

FIGURE 6. Mean (± SD) mass change by caged Red-eyed Vireos and *Catharus* thrushes fed four-day, ad libitum diets of insects (*Tenebrio* larvae), fruits (*Viburnum recognitum, Phytolacca americana,* and *Pyrus melanocarpa*) or a control diet of both insect and fruit resources. Preliminary experiments in 1993 included only exclusive fruit and insect diets.

logical processing. Thus birds may change food intake quantities, increase assimilation efficiency, and seek those fruits that allow them to maintain or gain weight on predominantly fruit diets, strengthening the argument that inclusion of fruit in the diet may benefit some species by providing greater energy intake during migration.

I have experimentally tested this idea that fruit is included in the diet to maximize energy intake by conducting feeding trials with transient Red-eyed Vireos and *Catharus* thrushes subjected to four-day treatments of fruit, insect, or mixed diets as in the migratory activity experiments above (Parrish 1996). In order to isolate energy intake from energy expenditure, I conducted feeding experiments in small holding cages (28 cm on all sides) that minimized energy expended by experimental individuals. In experiments conducted in 1993, *Catharus* thrushes (N = 21) gained significantly more mass on insect diets than on fruit diets, but did maintain or gain mass slightly on exclusively fruit diets (ANOVA; treatment × genus effect: $F = 26.8$, $P < 0.01$; Fig. 6), supporting the hypothesis that fruit can provide some energy intake benefits during stopover. Red-eyed Vireos, in contrast, always lost mass on fruit diets. In 1994, I added an additional treatment of mixed, ad libitum fruits and insects to experiments. In 1994, *Catharus* thrushes (N = 24) on average gained weight on mixed diets and insect diets. As in 1993, Red-eyed Vireos tested in 1994 (N = 29) also declined in energetic condition when provided with strictly fruit diets (ANOVA; treatment × genus interaction: $F = 3.75$, $P = 0.026$; Fig. 6), despite high levels of frugivory on the same fruits in the wild (Table 2). Mass gain did occur, however, when individual vireos fed on either insect and mixed control diets. A significant Bonferroni post hoc test demonstrated greater mass gain on average in mixed diets, where both fruit and insects were available ad libitum, compared to diet treatments of ad libitum insects alone ($P = 0.017$), suggesting a potential adaptive advantage to mixing diets by way of dietary plasticity during autumn migration (e.g., Pennings et al. 1993, Bairlein 1990; Fig. 6). These results with long-distance nearctic-neotropical migrants contrast with Berthold's (1976a,b) conclusions from the Palearctic, which suggest no increased energetic intake is possible for migrants on mixed or frugivorous diets. It is unknown at this time precisely what factors are responsible for the maintenance of extensive frugivory in Red-eyed Vireos in the wild, since experimental evidence suggests that large amounts of mass gain do not seem possible on exclusively fruit diets. However, the discrepancy between field patterns and feeding trial experiments may lie, at least in part, in the energy expended by Red-eyed Vireos when foraging on insect versus fruit diets.

Because of the differences between fruit and insect resources in the distribution, ease of capture, and rate of decline during the autumn migration period (Table 1, Fig. 2), I suggest that birds foraging on fruit expend significantly less energy during stopover foraging than those feeding exclusively on insects. Fruits are stationary, clumped, highly visible, and more predictably available than insects throughout autumn. These characters enable frugivorous migrants to remain virtually stationary in a patch of fruiting shrubs (Parrish 1996; Fig. 3) where they are limited in energy intake only by competition (Sealy 1989, Moore and Yong 1991) or by morphological or physiological constraints such as gut passage time or digestion and assimilation capabilities (e.g., Jordano 1987, Levey and Karasov 1989, Levey and Duke 1992).

I suggest that selection for minimization of energy expenditure during stopover has influenced the evolution of dietary plasticity during migration. Diet shifts to fruit allow many migrant species to minimize the time and energy needed for foraging on "expensive" diet types such as insects. The different energy requirements for foraging on fruit and insect resources may, in effect, alter the relative profitabilities of the two diet types, thereby influencing the proximate foraging decisions of migrant landbirds during stopover. Even for species which are incapable of efficiently assimilating exclusive diets of northern temperate zone fruits into lipid reserves (e.g., Red-eyed Vireos), bouts of insectivory for lipogenesis need only be infrequent and of short duration when caloric intake is subsidized by ingestion of easily acquired fruit resources. Preliminary feeding experiments with Red-eyed Vireos on limited and ad libitum treatments of insect diets support the idea that only small quantities of insects are necessary to effect significant weight gains and increases in fat loads (see also Izhaki and Safriel 1989). The reduction in energy expenditure during stopover foraging through dietary shifts to fruit can, in theory, be as important in effecting positive net energy balances during stopover periods as is maximization of energy intake through hyperphagia. Although the foraging observations reported herein suggest important energy savings during foraging on fruit, in practice, no study to date has tested experimentally the hypothesis that frugivory during migration significantly reduces the cost of stopover foraging relative to strictly insectivorous diets. A combination of energy expenditure reduction and an increase in energy intake and assimilation suggests that dietary plasticity actively facilitates successful stopover refueling for migrating landbirds.

IMPLICATIONS OF BEHAVIORAL PLASTICITY FOR THE CONSERVATION OF NEOTROPICAL MIGRANTS DURING *EN ROUTE* PERIODS

Because it affects aspects of migratory behavior that are important for a successful migration, such as habitat selection, foraging performance, and energetic budgets, dietary plasticity is clearly a phenomenon of direct importance to the conservation of migrant landbird populations. Shifts in diet can generate new resource needs, different energetic priorities, and novel behaviors for *en route* migrants that are entirely different from those used by the same species on the breeding or wintering grounds. Birds that are plastic in their diets will, as a result of their new resources, show changes in behavior that have important consequences for reversing declines of migrant populations. For example, changes that occur in *en route* habitat selection as a result of diet shifts can completely shift priorities for protection of migratory bird habitats that are based only on knowledge of breeding and wintering season biology. That certain thrush species breed principally in extensive hardwood and mixed coniferous-deciduous forests and winter in extensive tropical forest would certainly mask patterns of habitat use during migration through the northern temperate zone, when many species are highly abundant in small-statured coastal fruiting scrub (Parrish 1996, unpubl. data). Because of behavioral plasticity in the life cycles of migratory bird species, efforts to maintain the health and existence of desirable stopover habitat during migration require an understanding of transient migrants' needs during migration that is independent of breeding or wintering season behavior.

Furthermore, as the effects of dietary plasticity on habitat selection and use during migration continue to be illuminated, the present and future availability and vulnerability of these important stopover habitats must be considered in conservation decisions for migratory landbirds. For example, coastal habitats have long been considered important for songbird migration as a result of geographic and wind patterns that concentrate large numbers of species and individuals along coastal areas (e.g., Moore and Simons 1992a). Furthermore, coastal habitats are the areas most heavily used by hatching-year birds (perhaps due to navigational errors or inexperience; Ralph 1971, 1981), which annually recruit into the reproductive population and therefore are important first steps for recovery from declines. The evidence presented above suggests that fruit resources, which tend to be extremely common in coastal shrubland habitats, are very important for meeting the energy demands necessary for a successful migration.

Yet coastal habitats, especially along the eastern and southern seaboards of the United States, face unrelenting pressures for development, resulting from urban expansion, tourism, agriculture, and non-random population growth in coastal vicinities (Cullitan et al. 1990, Mabey and Watts *this volume,* Simons et al. *this volume*). Such degradation and destruction of migratory corridors along coastal North America represents a significant threat to migrating landbirds. Moreover, popular perception of shrubland habitats among coastal human communities is generally low (due to the high abundance in these habitats of thorny plants and poison ivy, *Rhus radicans*), resulting in land use matrices with little remaining of the fruit-bearing shrubland preferred by many migrating songbirds. Protection of existing maritime shrubland habitats and stewardship efforts aimed at managing for successional stages typified by an abundance of fruiting plants are encouraged given the importance of fruit in the diets and behaviors of many species. This example of fruit-bearing coastal shrubland demonstrates that the relative conservation value of habitats for landbirds is also a dynamic character, altered by seasonal dietary plasticity in the foraging behavior or nearctic-neotropical migrants.

Dietary plasticity, exemplified here as dietary shifts from insectivory to frugivory, illuminates an important, yet disturbing conclusion: *en route* migratory landbirds may possess completely different biologies than those with which we are familiar at other times of the year. Autumn and spring migrations are more than simple short-term links between reproduction and winter maintenance. They present additional, novel challenges to conservation biologists concerned with holistic approaches to migratory bird conservation. Until we recognize the patterns, causes, and consequences of seasonal changes, such as diet shifts, in the biological identity of landbird migrants, we will compromise any management efforts exerted during other periods of the annual cycle. Unbalanced approaches to conservation of migratory landbirds that are biased toward the breeding or wintering period may provide increased reproductive success or overwinter survival, benefits that could be lost in temporal population "sinks" during migration. Consideration of a broader view of nearctic-neotropical migrants (Levey 1994) that attempts to avoid temperate biases and accepts the changing biologies of these species throughout the annual cycle will be critical for migrant landbird conservation given the political, cultural, and biological boundaries over which these landbirds cross.

ACKNOWLEDGMENTS

I thank the following for their dedication, assistance, and companionship in the field: S. Comings, M. Whitman, C. Fuller, M. Grady, G. Lowe, S. Oppenheimer, A. Panjabi, A. Parker, A. Parrish, G. Turano, A. Williams, D. Workman, and numerous Earthwatch Research Corps volunteers. Mrs. Elise and Helen Lapham graciously allowed use of unpublished data and access to the field site. J. Clark and F. Moore allowed use of spring migration frugivory data to demonstrate the extent of dietary plasticity. Experimental cages were provided by R. L. Holberton and K. P. Able. Logistical support was generously provided by Mr. and Mrs. William Comings, Mr. and Mrs. Larry Smith, Mr. and Mrs. A. Angelbeck, Ms. Maureen Gary, and the Block Island Field Office of The Nature Conservancy. N. Wheelwright revealed the value of working with *Catharus* thrushes, D. Levey and J. Zurovchak provided stimulating discussions, and T. Kunz provided generous use of supplies, laboratory space, and ideas on experimental design with doubly labeled water techniques. D. H. Morse, F. R. Moore, R. L. Holberton, J. Rotenberry, and the Migratory Bird Group at the University of Southern Mississippi provided valuable comments on early drafts of the manuscript. Travel to the Stopover Ecology Symposium in Missoula, MT, was provided by a student travel award from the American Ornithologists Union. Field work was aided by a Earthwatch Research Grant, and a pre-doctoral fellowship and Dissertation Improvement Grant (IBN-9411650) to the author from the National Science Foundation. Sincere appreciation is extended to Frank Moore for the opportunity to take part in this migration symposium. This paper is dedicated to the memory of Mr. Peter Tortorigi.

APPENDIX. SPECIES OF NEARCTIC-NEOTROPICAL MIGRATORY LANDBIRDS KNOWN TO DEMONSTRATE FRUGIVORY DURING THE NON-BREEDING PERIOD

	Extent of Frugivory[a]			References		
Species	Autumn migration	Tropic winter	Spring migration	Autumn migration	Tropic winter	Spring migration
Yellow-billed Cuckoo *Coccyzus americanus*	+	+	++	10[b]	9	12[b]
Black-billed Cuckoo *Coccyzus erythropthalmus*	0	+		10[b]	9	
Olive-sided Flycatcher *Contopus borealis*		0			1, 9, 11, 24	
Eastern Wood Pewee *Contopus virens*	+	+/0	++	13	1, 2, 9, 11	12[b]
Western Wood Pewee *Contopus sordidulus*			+/0			9/11, 24
Eastern phoebe *Sayornis phoebe*	++	+		10[b], 16	9, 11, 13	
Say's phoebe *Sayornis saya*		0			9, 11	
Gray Flycatcher *Empidonax wrightii*		+/0			9/11	
Dusky Flycatcher *Empidonax oberholseri*		+/0			9/11	
Hammond's Flycatcher *Empidonax hammondi*		+/0			9/11	
Least Flycatcher *Empidonax minmus*	++	+		10[b]	9, 22	
Acadian Flycatcher *Empidonax virescens*	++	++	+	10[b]	9, 11, 24, 25[b]	12[b]
Traill's Flycatcher *Empidonax traillii*	++	++		10[b]	9, 11, 25[b]	
Ader Flycatcher *Empidonax alnorum*		+			1, 9, 11	
Yellow-bellied Flycatcher *Empidonax flaviventris*	++	+	++	10[b]	9, 11	12[b]
Western Flycatcher *Empidonax difficilis*		+/0			9/11	
Ash-throated flycatcher *Myiarchus cinerascens*		+/0			9/11	
Great crested Flycatcher *Myiarchus crinitus*	++	++	+	10[b]	8, 9, 11, 25[b]	12[b]

APPENDIX. CONTINUED

Species	Extent of Frugivory[a]			References		
	Autumn migration	Tropic winter	Spring migration	Autumn migration	Tropic winter	Spring migration
Scissor-tailed Flycatcher *Tyrannus forficatus*		++			9, 11	
Cassin's Kingbird *Tyrannus vociferans*		+/0			9/11	
Western Kingbird *Tyrannus verticalis*		+			9, 11	
Eastern Kingbird *Tyrannus tyrannus*		++	++	16	1, 5, 9, 11, 25[b]	12[b]
Tree Swallow *Tachycineta bicolor*	++	+	+	13	9	13
Violet-green swallow *Tachycineta thalassina*		0			9	
Purple Martin *Progne subis*		0			9	
Bank Swallow *Hirundo rustica*		0			9	
Northern Rough-winged Swallow *Stelgidopteryz serripennis*		0			9	
Cliff Swallow *Hirundo pyrrhonata*		+			9	
Barn Swallow *Riparia riparia*		0			9	
House Wren *Troglodytes aedon*	++	0	++	10[b]	9	12[b]
Marsh Wren *Cistothorus palustris*		0			9	
Sedge Wren *Cistothorus platensis*		0			9	
Winter Wren *Troglodytes troglodytes*	0	0		10[b]	9	
Ruby-crowned Kinglet *Regulus calendula*	+	+	0	10[b]	9	12[b]
Blue-gray Gnatcatcher *Polioptila caerula*		0	0		9	12[b]
Eastern Bluebird *Sialia sialis*	++	+		16, 20	3, 4, 9	
Western Bluebird *Sialia mexicana*		+			9	
Mountain Bluebird *Sialia currucoides*		+			9	
Townsend's Solitaire *Myadestes townsendi*		+			9	
Wood Thrush *Hyocichla mustelina*	++	++	++	10[b], 13, 14, 15, 16, 18[b], 20, 21	5, 7, 9, 11, 25[b]	12[b], 25[b]
Veery *Catharus fuscescens*	++	++	++	10[b], 13, 14, 16, 18[b], 20, 21	5, 9, 11, 25[b]	12[b]
Swainson's Thrush *Catharus ustulatus*	++	++	++	10[b], 13, 14, 15, 16, 18[b], 20, 21	5, 6, 7, 9, 11, 13, 17, 25[b]	12[b]
Gray-cheeked Thrush *Catharus minimus*	++	++	++	10[b], 13, 14, 16, 18[b] 20, 21	5, 9, 11, 25[b]	12[b]
Hermit Thrush *Catharus guttata*	++	++	++	10[b],13, 14, 15, 16, 18[b], 20, 21	9	12[b], 13
American Robin *Turdus migratorius*	++	++	++	10[b], 13, 14, 15, 16, 18[b], 20, 21	3, 4, 9, 11	13
Gray Catbird *Dumetella carolinensis*	++	++	++	10[b], 13, 14, 15, 16, 18[b], 20, 21	7, 9, 11, 25[b]	12[b]
Sage Thrasher *Oreoscoptes montanus*		+			9	
Cedar Waxwing *Bombycilla cedrorum*		++	++	10[b], 13, 16, 18[b], 20	9	12[b], 13

APPENDIX. CONTINUED

Species	Extent of Frugivory[a]			References		
	Autumn migration	Tropic winter	Spring migration	Autumn migration	Tropic winter	Spring migration
White-eyed Vireo *Vireo griseus*	++	++	++	10[b], 13	2, 7, 9, 11, 22	12[b], 13
Bell's Vireo *Vireo bellii*		+			9	
Black-capped Vireo *Vireo atricapillus*		+			9	
Yellow-throated Vireo *Vireo flavifrons*		+	++		9, 11, 22	12[b]
Gray Vireo *Vireo vicinior*		+			9	
Solitary Vireo *Vireo solitarius*	++	++		10[b], 13c	2, 9	
Warbling Vireo *Vireo gilvus*	++	+		10[b], 13, 16, 20	2, 9	
Red-eyed Vireo *Vireo olivaceus*	++	++	++	10[b], 13, 14, 15, 16	2, 9, 11, 23[b], 25[b]	12[b]
Philadelphia Vireo *Vireo philadelphicus*	++	+		10[b], 13, 15,	9	
Prothonotary Warbler *Protonotaria citrea*		++	+		5, 8, 9, 11, 23[b]	12[b]
Golden-winged Warbler *Vermivora chrysoptera*		0			9, 11	
Blue-winged Warbler *Vermivora pinus*		0/+	0		9, 11/25[b]	12[b]
Tennessee Warbler *Vermivora peregrina*	+	++	+	13	5, 8, 9, 11	12[b]
Orange-crowned Warbler *Vermivora celata*	+	0		13	9, 11	
Nashville Warbler *Vermivora ruficapilla*	++	0		10[b]	9, 11	
Virginia's Warbler *Vermivora virginiae*		0			9	
Lucy's Warbler *Vermivora luciae*		0			9	
Northern Parula *Parula americana*	++	+	+	10[b]	9, 22	12[b]
Black and White Warbler *Mniotilta varia*	+	0	+	10[b]	9, 11, 25[b]	12[b]
Black-throated Blue Warbler *Dendroica caerulescens*	++	++		10[b]	11, 13	
Cerulean Warbler *Dendroica cerulea*		0	+		9, 11	12[b]
Blackburnian Warbler *Dendroica fusca*	+	++		10[b]	6, 9, 25[b]	
Chestnut-sided Warbler *Dendroica pensylvanica*	++	+	+	10[b]	9, 11, 25[b]	12[b]
Cape May Warbler *Dendroica tigrina*	+	+	++	13	11, 19	12[b]
Magnolia Warbler *Dendroica magnolia*	++	+/0	+	10[b]	22/9, 25[b]c	12[b]
Yellow-rumped Warbler *Dendroica coronata*	++	++	++	10[b], 14, 16, 18[b], 20, 21	9, 11, 19	12[b], 13
Black-throated Gray Warbler *Dendroica nigrescens*		0			9	
Townsend's Warbler *Dendroica townsendi*		0			9	
Hermit Warbler *Dendroica occidentalis*		0			9	
Black-throated Green Warbler *Dendroica virens*	++	+	0	10[b]	9, 22	12[b]

APPENDIX. CONTINUED

Species	Extent of Frugivory[a]			References		
	Autumn migration	Tropic winter	Spring migration	Autumn migration	Tropic winter	Spring migration
Golden-cheeked Warbler *Dendroica chrysoparia*		0			9, 25[b]	
Yellow-throated Warbler *Dendroica dominica*		0			9	
Grace's Warbler *Dendroica graciae*		0			9	
Prairie Warbler *Dendroica discolor*		+	0		9	12[b]
Bay-breasted Warbler *Dendroica castanea*	+	++	+	13	9, 11, 19, 25[b]	12[b], 13
Blackpoll Warbler *Dendroica striata*	++	+	+	10[b]	9	12[b]
Pine Warbler *Dendroica pinus*		+			9	
Palm Warbler *Dendroica palmarum*	++	0		10[b]	9, 11	
Yellow Warbler *Dendroica petechia*	++	0	++	10[b]	9, 11	12[b], 13
Mourning Warbler *Oporornis philadelphia*	++	0		10[b]	9, 11, 25[b]	
MacGillivray's Warbler *Oporornis tolmiei*		0			9, 11	
Connecticut Warbler *Oporornis agilis*	+	0		10[b]	9	
Kentucky Warbler *Oporornis formosus*		0/+	+		11/9, 25[b]	12[b]
Canada Warbler *Wilsonia canadensis*	+	0		10[b]	9, 11, 25[b]	
Wilson's Warbler *Wilsonia pusilla*	+	0/+		10[b]	11/9	
Hooded Warbler *Wilsonia citrina*	+	0	+	10[b]	9, 11, 25[b]	12[b]
Worm-eating Warbler *Helmitheros vermivorus*		0/+	+		9, 11/25[b]	12[b]
Swainson's Warbler *Limnothlypis swainsonii*		?	+			12[b]
Ovenbird *Seiurus aurocapillus*	++	+	+	10[b]	11, 25[b]	12[b], 13
Louisiana Waterthrush *Seiurus motacilla*		0	+		9, 11, 25[b]	12[b]
Northern Waterthrush *Seiurus noveboracensis*	++	0	+	10[b]	9, 11, 25[b]	12[b]
Common Yellowthroat *Geothlypis trichas*	++	0	+	10[b]	11	12[b]
Yellow-breasted Chat *Icteria virens*	++	0/+	++	10[b], 20	11/9	12[b]
American Redstart *Setophaga ruticilla*	0	0	+	10[b]	9, 13	12[b]
Rose-breasted Grosbeak *Pheucticus ludovicianus*	++	++	++	10[b], 20,	3, 5, 9, 11, 13	12[b]
Black-headed Grosbeak *Pheucticus melanocephalus*		+			9	
Blue Grosbeak *Guiraca caeurula*		++	++		3, 9	12[b]
Indigo Bunting *Passerina cyanea*		++	++		3, 9	12[b]
Lazuli Bunting *Passerina amoena*		0			9	
Savannah Sparrow *Passerculus sandwichensis*		0			9	

APPENDIX. CONTINUED

Species	Extent of Frugivory[a]			References		
	Autumn migration	Tropic winter	Spring migration	Autumn migration	Tropic winter	Spring migration
Vesper Sparrow *Pooecetes gramineus*		0			9	
Lark Sparrow *Chondestes grammacus*		0			9	
Chipping Sparrow *Spizella passerina*	+	0		10[b]	9	
Clay-colored Sparrow *Spizella pallida*		0			9	
Brewer's Sparrow *Spizella breweri*		0			9	
White-crowned Sparrow *Zonotrichia leucophrys*	++	0	+	10[b], 15,	9	13
Lincoln's Sparrow *Melospiza lincolnii*	++	0		10[b]	9	
Bobolink *Dolichonyx oryzivorus*		0			9	
Orchard Oriole *Icterus spurius*		++	++	20	4, 5, 9,	12[b]
Northern Oriole *Icterus galbula*	++	++		10[b], 14, 20	4, 5, 9, 13	
Scarlet Tanager *Piranga olivacea*	++	++	++	10[b], 14, 16	3, 5, 6, 7, 9, 11, 25[b]	12[b]
Western Tanager *Piranga ludoviciana*		+			9	
Summer Tanager *Piranga rubra*	+	++	++	20	3, 5, 6, 9, 11, 25[b]	12[b]

Notes: Where there is strong disagreement among the literature sources, both scores of frugivory and their respective references are provided. Source reference list includes earlier studies that reviewed diets during the wintering period (9, 11, 19; see below) and complete reference lists for frugivory in landbird migrants while wintering in the tropics can be obtained there.

[a] Extent of frugivory determined by counts of the number of primary literature records (as in Levey and Stiles 1992) or based on categorization of data from quantitative studies. During spring migration, few studies have attempted to document frugivory, and the extent of frugivory is therefore based predominantly on an unpublished study from J. Clark and F. Moore from the Gulf coast. For autumn migration and for tropical wintering periods, the extent of frugivory was determined as follows: (0) = no reports of frugivory or 0–5% of samples in quantitative studies containing fruit; (+) = 2–5 different records of frugivory from the literature or 5–25% of samples in quantitative studies containing fruit; and (++) = >5 different records of frugivory from the literature or > 25% of samples in quantitative studies containing fruit.

[b] Quantitative study using diet/fecal analyses in which frequency of frugivorous samples was determined.

[c] Source references: (1) Fitzpatrick 1980; (2) Barlow 1980; (3) Faaborg 1980; (4) Hutto 1980; (5) Willis 1980; (6) Hilty 1980; (7) Rappole and Warner 1980; (8) Morton 1980; (9) Rappole et al. 1993; (10) Parrish 1996; (11) Levey and Stiles 1992; (12) J. Clark and F. Moore, unpubl. data; (13) J. Parrish, pers. obs.; (14) Blake and Hoppes 1986; (15) Rybczynski and Riker 1981; (16) Davidar and Morton 1986; (17) Howe 1981; (18) White and Stiles 1990; (19) Greenberg 1979; (20) Stiles 1980; (21) Malmborg and Willson 1988; (22) Greenberg et al. 1995a; (23) Poulin et al. 1994; (24) Sherry 1984; (25) Blake and Loiselle 1992a.

Studies in Avian Biology No. 20:71–87, 2000.

DISRUPTION AND RESTORATION OF *EN ROUTE* HABITAT, A CASE STUDY: THE CHENIER PLAIN

WYLIE C. BARROW, JR, CHAO-CHIEH CHEN, ROBERT B. HAMILTON, KEITH OUCHLEY, AND TERRY J. SPENGLER

Abstract. Cheniers (relict beach ridges) and other habitats adjacent to ecological barriers may be critical linkages in the migratory pathways of long-distance migratory birds. It is important that these wooded habitats provide enough food and cover at the right time to support these birds' requirements. To date, little attention has been given to the conservation of coastal woodlands, habitats in which *en route* migrants tend to concentrate in large numbers during migration. Because about one-third of North America's human population lives within 80 km of the coast, many forest-dwelling landbird migrants now depend on degraded native woodlands and urbanized environments for survival during migration. Restoration or rehabilitation of coastal woodlands, such as the cheniers of southwest Louisiana and southeast Texas, is of particular importance because of historic anthropogenic modifications, their limited geographic extent, and the extraordinary abundance and species richness of migratory birds using them during migration. In this paper, we use the Chenier Plain as a case study to discuss the issue of land use changes and their consequences for maintaining suitable stopover habitat. Results from an ongoing field study in this ecosystem indicate that most forest-dependent migratory birds are tolerant of at least some degradation of chenier forest during migration. However, these results reveal that subtle differences in vegetation composition and structure beneath the canopy of these forests, primarily as a result of livestock grazing and white-tailed deer overbrowsing, can result in differential use by some *en route* migrants. Species that were most affected by disturbance to the forest understory were early-arriving migrants, dead-leaf foragers, frugivores, and nectarivores. Given that the understory structure and regeneration of chenier forests has been so greatly reduced, and that high densities of nearctic-neotropical migrants tend to concentrate in cheniers during migration, restoration and rehabilitation should be conservation priorities in the Chenier Plain.

Key Words: Chenier Plain, habitat degradation, habitat restoration, habitat use *en route,* migration, nearctic-neotropical migrant landbirds, plant species use.

The increasing international attention given to the nearctic-neotropical migratory bird system usually focuses on the hemispheric implications of deforestation, such as the clearing of tropical forests (Briggs and Criswell 1978, Terborgh 1980, Lovejoy 1983, Hagan and Johnston 1992, Rappole and McDonald 1994) and the fragmentation of temperate forests (Whitcomb 1977, Whitcomb et al. 1981, Wilcove and Whitcomb 1983, Hagan and Johnston 1992). Little attention is given to the disruptive events that may occur along the migration routes at important staging or stopover sites (Moore et al. 1993, Parker 1994). Due to the rapid expansion of human populations in the Western Hemisphere (Meyer and Turner 1992, Bongaarts 1994), migratory birds encounter an increasingly degraded landscape throughout their migration pathway each year.

Because *en route* migrants tend to concentrate in habitats adjacent to ecological barriers, several key landscape features have been identified as being stopover sites or staging areas of special concern to forest-dwelling birds. These landscape features include forests on and adjacent to mountain ranges; woodland patches in and adjacent to agricultural, grassland, and urban landscapes; coastal hardwoods near large water bodies; and riparian vegetation and gallery forests in arid landscapes (Sprunt 1975, Moore et al. 1993). The degree to which the survival of long-distance migrants depends on strategically dispersed wooded stopover sites is unknown (Parker 1994). We do know, however, that these key stopover areas are precisely the locations that have received the greatest extent of anthropogenic modifications in the Americas, and are projected to be areas that will be receiving the greatest human population increases in the future. For example, coastal habitats (e.g., Atlantic and Gulf coasts) are known to be critically important stopover and staging areas (Sprunt 1975, McCann et al. 1993, Moore et al. 1993). In North America at present, about one-third of our population resides in coastal counties, and by the year 2010, as much as 75% of our population may live within 80 km of the coast (US Department of the Interior 1993). Nearctic-neotropical migrants now depend on degraded natural areas and urbanized environments for survival throughout their annual cycle (Morrison et al. 1994, Greenberg et al. 1995b). The fact that disturbed habitats will play an increasingly important role in the conservation of long-distance migrants accentuates the need for habitat restoration planning and implementation.

In this chapter we discuss the issue of land-use changes and their consequences, and we propose an approach to ameliorate rates of native woodland loss, namely the rehabilitation of damaged stopover sites and staging areas. Habitat rehabilitation provides a fundamental solution to current problems of woodland loss. We focus on the Chenier Plain of the Gulf of Mexico but believe that the principles described here apply to all key stopover areas.

We use the Chenier Plain as a case study to illustrate the various types of activities that produce damaged or degraded lands, provide results from an ongoing field study, discuss the ecological strategies for rehabilitating these lands, and conclude with a statement of directions for future work on the rehabilitation of *en route* habitat.

Conservation activities can be applied at several geographic scales. To be successful, restoration/rehabilitation of habitat should take place within the context of the conservation of an entire migration pathway. Unfortunately, exact migration routes (i.e., width, shape, orientation) for most species are unknown (Russell et al. 1994), especially those portions of the route that travel through the tropics (Parker 1994). At the next level of planning, decisions should be made within the context of the conservation of the entire landscape. Simons et al. (*this volume*) identify landscape-level factors to consider for conservation planning. We focus here on the local-level (i.e., within-habitat) features that may influence the suitability of a habitat to birds during migration.

HISTORY AND IMPORTANCE OF THE CHENIER PLAIN

Description of the Chenier Plain

The geographic limits of the Chenier Plain are the Bolivar Peninsula and East Bay, Texas, on the west; the Gulf of Mexico on the south; Vermillion Bay, Louisiana, on the east; and the transition from plain to prairie and longleaf pine flatwoods on the north. The Chenier Plain comprises about 322 km of coastal landforms that lie perpendicular to the path of the trans-Gulf bird migration, and extends inland by distances ranging from 16 km to 64 km; total area is about 1,295 km^2 (Gosselink et al. 1979). The land area is approximately 750,000 ha, exclusive of water, and is comprised mostly of wetland habitats (i.e., marsh and coastal prairie). Wooded habitats occur as upland forest on salt domes and Pleistocene islands (1.2%); coastal hardwoods on relict beach ridges (i.e., cheniers), man-made levees, and spoil banks (4.5%); and bottomland hardwood and swamp forest along river systems (0.5%) (Gosselink et al. 1979).

The term "chenier" is French meaning "place of oak", referring to the upland areas occurring in the marshland of the Chenier Plain that are typically covered with a coastal live oak (*Quercus virginiana*) plant community. These ridges are of three basic origins: barrier islands, river mouth accretions, and relict beach ridges. The origins of the cheniers are explicable in terms of the fluctuation in Mississippi River delta sedimentation. Repeated occurrence of heavy sediment influx as the Mississippi River advanced westward, followed by lapses in the supply as the Mississippi River retreated eastward, created a series of cheniers as the shoreline periodically advanced gulfward. To a lesser extent, the Sabine, Calcasieu, and Mermantau rivers are contributors of sediment during chenier ridge formation (Taylor et al. 1995). Fluctuations of mud and sand supply to this region of the Gulf coast may also reflect periods of high and low floods in the Mississippi drainage, as well as delta shifts (Spearing 1995). Cheniers lie landward of primary beach dunes, up to a distance of about 20 km inland from the coast (Fig. 1). Ridges of southwest Louisiana generally trend east-west and have similar alignment as the present shoreline; some ridges have multiple crests and swales (Taylor et al. 1995). They range from 30–500 m in width, from a few centimeters to more than 3 m in elevation, and may extend coastwide for distances of 55 km or more without interruption (Russell and Howe 1935, Byrne et al. 1959).

In Louisiana, the associated plant community in its natural condition is most often a forest dominated by live oak and hackberry (*Celtis laevigata*). Chenier ridges in Texas, however, are often grasslands or shrub thickets that only locally support oak vegetation (Texas Bureau of Economic Geology 1976). According to Cocks (1904, 1907), Palmisano (1970), and W. Barrow (pers. obs.), other woody plant species that typically coexist on these ridges include water oak (*Quercus nigra*), red mulberry (*Morus rubra*), toothache-tree (*Zanthoxylum clava-herculis*), pecan (*Carya illinoensis*), green ash (*Fraxinus pennsylvanica*), common persimmon (*Diospyros virginiana*), bumelia (*Bumelia lanuginosa*), Chinese tallow (*Sapium sebiferum*), and honeylocust (*Gleditsia triacanthos*). The understory consists primarily of deciduous holly (*Ilex decidua*), yaupon (*Ilex vomitoria*), cherry laurel (*Prunus caroliniana*), sweet acacia (*Acacia farnesiana*), blackberry (*Rubus* sp.), swamp dogwood (*Cornus drummondii*), green hawthorn (*Crataegus viridis*), palmetto (*Sabal minor*), and Carolina wolfberry (*Lycium carolinianum*), and numerous vines such as greenbrier (*Smilax* spp.), grape vine (*Vitis cinerea*), Carolina moonseed

FIGURE 1. Aerial view of Hackberry Ridge, a forested chenier complex (relict beach ridges), in Cameron Parish, LA. Hackberry (*Celtis laevigata*) is the dominant canopy tree. Cattle ranching is the primary land use practice.

(*Cocculus carolinus*), poison ivy (*Toxicodendron radicans*), rattan-vine (*Berchemia scandens*), trumpet-creeper (*Bignonia radicans*), virginia creeper (*Parthenocissus quinquefolia*), and ladies eardrop-vine (*Brunnichia cirrhosa*). Epiphytes such as Spanish moss (*Tillandsia usneoides*), resurrection fern (*Polypodium polypodioides*), and numerous species of mosses (e.g., *Cryphaea* spp. and *Thuidium* spp.) and lichens (e.g., *Usnea florida* and *Physcia* spp.) are patchily distributed and can be locally common (Cocks 1907, Reese 1984). As in other maritime forests, much of the plant species diversity is a result of the woody, smaller-statured species of the understory.

Importance of the Chenier Plain as a Stopover Area for Nearctic-neotropical Migrants

Historically, the landscape of this region has been recognized as one of the most important physiographic areas to migratory birds in North America (Gauthreaux 1971, Lowery 1974, Sprunt 1975, Bellrose 1976, Gosselink et al. 1979, Moore et al. 1993). Although forests were not originally extensive in the Chenier Plain, by reason of their geographic position they play a key role in the nearctic-neotropical bird migration system. At least 63 species of migratory birds regularly use these wooded habitats prior to, or immediately after, crossing the Gulf of Mexico. The spring-migration period in the Chenier Plain extends from late February through May with peak numbers occurring between mid-April and early May (Fig. 2). Migrants are found in coastal forests somewhat erratically in March and then on a regular, almost daily basis in April and the first half of May (Gauthreaux 1971, Lowery 1974).

In spite of the difficulty in crossing an ecological barrier as large as the Gulf of Mexico, many trans-Gulf migrants continue flight inland and make landfall to the north of the Chenier Plain (Lowery 1945, Gauthreaux 1971). Intuitively, one would not expect the evolutionary strategies of trans-Gulf migrants to develop such that migrants would have to land along the coast since this would leave no room for error if the flight becomes difficult. The greater expanses of forest found farther inland would also appear to provide more suitable habitat for *en route* migrants than is available in the Chenier Plain. The "coastal hiatus" (Lowery 1945, 1951) and the days of few migrants encountered in the chen-

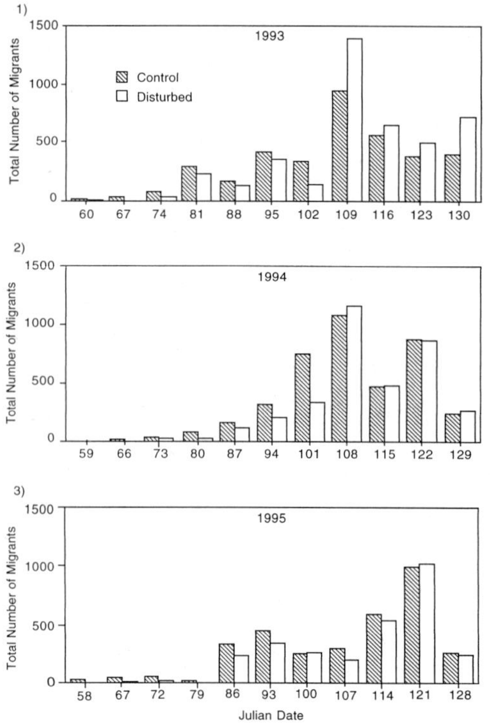

FIGURE 2. Comparison of total number of nearctic-neotropical migrants using control and disturbed plots during 12 weeks across three study sites in the Chenier Plain during 1993, 1994, and 1995.

iers, plus the few repeat birds between years (W. Barrow, unpubl. data), suggest that during the spring-migration period cheniers are crucial only at certain, rather infrequent occasions, and then only for individuals. However, due to the restricted spring-migration periods of some migratory species through the Chenier Plain (Lowery 1974; W. Barrow, unpubl. data), the infrequent occasions that migrants depend on cheniers could potentially impact large numbers of individuals of declining species. Furthermore, cheniers may be used to a greater extent by fall migrants, which travel southwest across the Chenier Plain (Able 1972), perhaps using cheniers as habitat corridors.

We can begin to determine the importance of cheniers as stopover habitat from radar technology, which provides the percentage of *en route* trans-Gulf migrants using cheniers. Gauthreaux (1971) revealed that during advantageous weather conditions (fair skies and southerly winds) about 10% of migrants flying across the Gulf of Mexico in spring land in the Chenier Plain at locations south of Lake Charles, Louisiana. However, during adverse weather conditions (hard rain and/or northerly winds), as many as 80% of individuals in a trans-Gulf flight alight in the wooded uplands of this region. Able (1972), also using radar facilities in Lake Charles, Louisiana, noted that trans-Gulf flights were less frequent during autumn migration, but the Chenier Plain was heavily used by birds moving southwest along the Louisiana and Texas coasts. These southwest flights across the Chenier Plain result in concentrations of fall migrants in chenier habitats that are similar to those observed during unfavorable weather conditions during spring migration (W. Barrow, unpubl. data).

Because of events that are occurring in the tropics, namely human population growth and subsequent development of the Gulf coastal plain, the Chenier Plain will likely play an even greater role in the conservation of the nearctic-neotropical migration system. Consider the following: nearctic-neotropical migrants need to accumulate large amounts of fat prior to crossing the Gulf, or any large ecological barrier. During the next decade, staging areas in Central America and Mexico will face an increasingly greater risk of degradation (Parker 1994). As key staging areas are disrupted, the effective width of the Gulf of Mexico will increase as a consequence of these land-use changes along the southern margins of the Gulf or at sites as yet unidentified. If migrants are prevented from depositing adequate fat stores prior to Gulf crossing, cheniers and human-created habitats may provide increasingly important feeding stations *en route*.

Historical Changes in Habitat Structure and Composition

Habitats within the Chenier Plain have gained or lost area in response to natural processes, human-induced disturbances, or both. Wind and wave disturbance from hurricanes can reduce chenier habitat, especially when the vegetative cover has been altered. In addition, the forces that created the cheniers of Louisiana and Texas (sediment influxes and lapses due to a shifting Mississippi River delta) are no longer in effect today. Control structures were built during the 1950s to prevent further diversion of the river's main outlet and the subsequent westerly discharge of sediment. Without these control structures, it has been predicted that the Mississippi River would have changed its course to flow through the Atchafalaya River approximately twenty years ago (US Department of the Interior 1978). In addition, the net subsidence rate for the land in the Chenier Plain is about 1.7 cm per year (Gosselink et al. 1979). Subsidence of cheniers will likely proceed until their burial under marsh deposits becomes general everywhere

FIGURE 3. Aerial view of Peveto Woods, a chenier forest remnant, in Cameron Parish, LA. Disruptive land use practices evident on this chenier are private homesteads, sand mining pits, an industrial plant, highway and road development, and conversion of forest cover to pasture. Invasive exotic plants and coastal erosion are also a concern on this chenier.

to the north of coastal beaches. The exposure of water cover in the Chenier Plain increased 35% from 1956 to 1988 (National Wetlands Research Center Map 93-02-027). The subsidence and erosion normally associated with the northern Gulf coast are thus no longer being offset by the influx of sediment from the Mississippi River. Coastal lands, including cheniers, are or will be disappearing as long as subsidence and erosion continue to prevail along the northern Gulf coast.

Many recent changes in the Chenier Plain landscape have resulted directly from the influence of human settlement and exploitation (Fig. 3). Because cheniers are the only well-drained ground in the region, these areas are the most suitable for human development. The cheniers of Louisiana and Texas have been occupied by Europeans for over 200 years, and consequently, the factors that have affected the suitability of these habitats to nearctic-neotropical migrants are numerous. Some examples include residential and industrial development, recreational camps, conversion to croplands and pasture, oil and gas exploration and development (canals, levees, and spoil banks), water control structures, controlled burning, white-tailed deer (*Odocoileus virginianus*) over-browsing (indirectly caused by humans), feral goat and cattle grazing, mosquito abatement programs, highway development, invasion of exotic plants, and coastal subsidence and sea-level rise.

The cumulative impact of the above factors have resulted in a disruption of the wooded habitat that originally occurred on the cheniers. Significant portions of most original chenier habitat have been cleared for agriculture (cotton), cattle production, and human settlement since the 1800s. Cattle ranchers prefer some tree cover for shade, although the understory vegetation remains altered. More recent losses are related to the population growth that is occurring throughout the Gulf coast region. Between 1952 and 1974 in the Chenier Plain, natural chenier forest, upland forest (primarily on salt domes), and swamp forest declined by approximately 1,250 ha, 1,250 ha, and 396 ha, respectively. Cropland (primarily rice) and urban areas had a net increase of 10,059 ha during the same period. Inland open water increased by 28,026 ha, representing the largest net area change during these 23 years (Gosselink et al. 1979). Continued expansion of the human population in this area will probably occur at the expense of the few re-

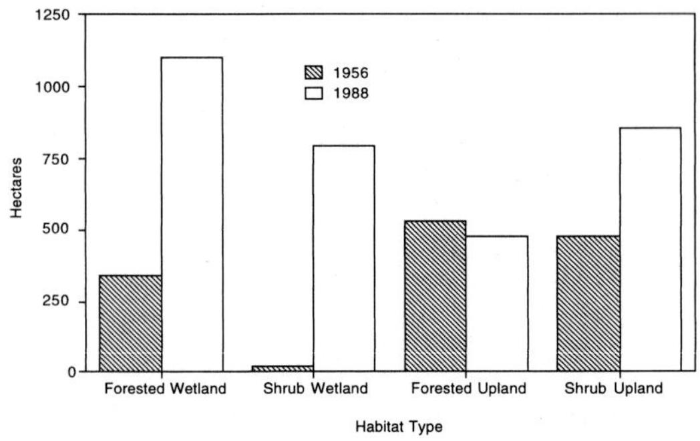

FIGURE 4. Changes in wooded habitats from 1956 to 1988 in the Chenier Plain. Data are from National Wetlands Research Center Map 93-02-027.

maining chenier forests and upland agricultural areas.

Despite the loss of chenier forests, there has been an overall increase in woody vegetation since 1956 (Fig. 4). This increase in woody vegetation can be primarily attributed to the creation of human-made levees and spoil banks that support woody vegetation and the natural succession of abandoned agricultural fields. Canals form an extensive network throughout the marshland of the Chenier Plain. Levees and spoil banks (20–75 m in width) associated with these canals support upland vegetation previously absent from the coastal marsh landscape. The total length of canal networks in the Chenier Plain is greater than 8,715 km (Gosselink et al. 1979). As of 1979, these habitats already occupied more than 2% of the total area of the Chenier Plain, an area greater than that of the beach, swamp-forest, upland-forest, and salt-marsh habitats combined (Gosselink et al. 1979). The vegetation on the levees and spoil banks is largely comprised of an exotic species, Chinese tallow. The shrub (dominated by sapling tallow trees) and forested wetland (dominated by older tallow tree thickets) categories in Fig. 4 illustrate the rapid expansion of this exotic plant during the past 40 years.

Of particular interest is the ability of Chinese tallow to invade the coastal prairie even though native trees in the area are restricted to relict beach ridges and riparian sites. The greater ability to survive occasional droughts on the heavy clay soil of the Chenier Plain is thought to be an important factor to its successful invasion (Bruce 1993). Tallow trees can facilitate the creation of new woodlands (Bruce et al. 1995). Bruce (1993) found that tallow trees improve germination conditions for other woody plant species in the coastal prairie. Depending on the age of the woodland, the understory and co-dominates of tallow woodlands in the Chenier Plain typically consist of dense stands of tallow saplings, wax myrtle (*Myrica cerifera*), yaupon, hackberry, oak species, groundsel bush (*Baccaharis halimifolia*), and various species of vines. It remains uncertain whether colonization of tallow woodlands by native species will eventually replace tallow, or if tallow will retain dominance (Bruce et al. 1995).

The remaining "natural" coastal forest is suffering from reduced overstory regeneration, elimination of understory vegetation, altered plant species composition, and spread of exotic plants. This is, in part, the result of cattle ranching. Essentially all (>95%) cheniers in Louisiana are grazed by cattle (M. Mattox, Soil Conservation Service, pers. comm.), causing the forested landscape on cheniers to be comprised of a structurally non-diverse mosaic of forested habitats that is park-like in appearance. Although cattle density and grazing schedules vary among cheniers, the general ranching system that has developed over the past two centuries is one of seasonal movement of cattle between cheniers/marshland and the coastal prairies to the north. Cattle graze cheniers and adjacent marsh from about 15 October through 15 May. The cattle are then transported to the coastal prairies north of the cheniers and marshland to graze during the summer months, where insects are less of a problem. Cattle movement among the cheniers is facilitated by more than 400 km of cattle walkways that traverse marshland and connect or nearly connect the isolated, parallel ridges. These walkways are small earthen levees about 1–3 m in elevation that were constructed across marshland during the 1920s and continue to be maintained today (Gosselink et al. 1979). Cattle grazing and pasture development have been shown to negatively affect breeding (Mosconi and Hutto 1983, Taylor 1986, Taylor and Littlefield 1986; J. Rappole, unpubl. report to USFWS), *en route* (Mueller and Sears 1987), and wintering nearctic-neotropical migrants (Saab and Petit 1992). However, in the absence of American Bison (*Bison bison*; Newcomb 1961), cattle may be instrumental in maintaining the dispersal and distribution (through differential grazing) of certain plant species important to migrants, such as honeylocust and *Acacia* spp. (Vines 1960, Fowells 1965).

A natural disturbance that may influence plant distribution is fire. Although not well-documented, chenier forests may have been subject to frequent fires. Fire played an important role in the maintenance of the adjacent plant communities, coastal prairie and marshland. How alteration of the historic fire regime, by use of controlled burning programs, has influenced plant species composition in chenier forests is not known to us.

On the more heavily disturbed cheniers, plant species composition usually shifts to a habitat dominated by nonindigenous species. Examples include chickasaw plum (*Prunus angustifolia*), cherokee (*Rosa laevigata*) and McCartney roses (*Rosa bracteata*), prickly pear cactus (*Opuntia lindheimeri*), chinaberry (*Melia azedarach*), Japanese honeysuckle (*Lonicera japonica*), privet (*Ligustrum* spp.), Chinese wisteria (*Wisteria sinensis*), lantana (*Lantana camara*), salt cedar (*Tamarix sallica*), and Chinese tallow (Cocks 1904, Palmisano 1967; W. Barrow, pers. obs.). All of these species are exotic except prickly pear cactus and chickasaw plum. How the encroachment of these exotic plants have affected the native plant species is unknown.

Restoration efforts to create historic plant communities in the Chenier Plain may be impractical because we are not even certain of the true pre-settlement plant species composition and relative abundance. A review of the trees recorded in the original land surveys of the Chenier Plain needs to be investigated (see Siccama 1971, Delcourt and Delcourt 1974). One species that apparently was common in the past is the toothache-tree (Hine 1906, Cocks 1907, Billings 1909). Hine (1906:68) considered this species "a common tree that grows in the locality" (referring to the Chenier Plain). Billings (1909:1) described a chenier in Cameron Parish, Louisiana, as "partially covered with a growth of stunted trees, principally *Xanthoxylum clava-herculis.*" Cocks (1907:6) described the region of the Chenier Plain in the vicinity of the Calcasieu river as "practically treeless, with the exception of some thickets of *Bumelia lanuginosa* and *Xanthoxylon clava-herculis,* and a few stunted specimens of hackberry on the ridges." The toothache-tree may have been a "keystone species" in the Chenier Plain. For instance, Hine (1906:68) described an insect outbreak specific to this plant: "a scale insect, *Ceroplastes cirripediformis,* is abundant on this tree, in fact, some trees are literally covered with it. The scale is rather large and must secrete a great deal of honeydew, for insects of different orders were attracted in large numbers. Many species not seen anywhere else were plentiful enough here to make them appear common." The large and diverse insect community associated with the toothache-tree would make it quite valuable to insectivorous birds during migration (Hine 1906). Many *en route* migrants also eat the fruit of toothache-trees in the fall (Vines 1960). An investigation into the apparent demise of this plant species seems warranted.

MICROHABITAT USE BY *EN ROUTE* MIGRANTS

Extrinsic benefits, such as wind conditions and location along evolutionarily programmed migration routes, will initially determine the use of some stopover habitats (Hutto 1985b, Moore and Simons 1992a, Moore et al. 1993). Although numerous studies have examined habitat use and foraging ecology of nearctic-neotropical migrants on their temperate breeding grounds, relatively few studies have examined the patterns of resource exploitation during migration. The quality of stopover habitats affects the physiological, behavioral, and population ecology of Rufous Hummingbirds (*Selasphorus rufus*), and perhaps other species as well (Russell et al. 1994). Selections between habitats at stopover sites have largely been attributed to food availability (Bibby et al. 1976, Martin 1980, Bairlein 1983, Bibby and Green 1983, Graber and Graber 1983; Hutto 1985a,b; Martin 1985, Lindström 1990b, Winker et al. 1992a). Previous *en route* studies suggest that migrants select among habitats during stopover (Hutto 1981, 1985a,b; Moore et al. 1990, 1993; Moore and Simons 1992a, Winker et al. 1992a), especially for structurally complex habitats (Moore et al. 1993). Parnell (1969) reported that several species of migrants in North Carolina during spring migration exploit sites similar to those of their breeding grounds. Parnell (1969) concluded that, for certain species, the selection of certain areas within a habitat (microhabitat use) might be more important than habitat choice.

Within-habitat selection during stopover is poorly understood (Moore et al. 1993), but food availability appears to also determine microhabitat use (Hutto 1985b). Graber and Graber (1983) studied foraging movements and food availability of spring-migrant warblers in Illinois. They found that differences in migrants' foraging patterns depended on the abundance of available arthropods. At stopover sites with low prey abundance, migrants foraged rapidly and were unsuccessful in finding enough food to permit weight gain. Most birds departed low-prey sites after one day. It may be critical for migratory birds to find one or more stopover sites with high prey availability along the migratory pathway (Graber and Graber 1983, Myers et al. 1987).

Loria and Moore (1990) demonstrated that fat-depleted Red-eyed Vireos (see Appendix for scientific names of all birds) in the Chenier Plain following trans-Gulf migration diversified their foraging behavior and expanded their use of microhabitat space. In another study at the same site, Moore and Yong (1991) used a predator-exclosure experiment to demonstrate that trans-Gulf migrants can depress prey abundance during stopover in chenier forests. Migrants foraging during periods of high migrant abundance were not able to replenish energy reserves as rapidly as migrants using the woodlands during periods of low migrant abundance (Loria and Moore 1990). If species that typically forage in the understory are forced to shift to the canopy or ground for food because of understory alterations, how does the increased abundance of potential competitors in these different microhabitats affect their ability to replenish energy reserves?

FIELD STUDIES OF HABITAT USE IN CHENIER FORESTS

We studied *en route* landbird migrants at three locations situated approximately 60 km apart in the Chenier Plain: (1) Grand Chenier, Cameron Parish, Louisiana; (2) Hackberry Ridge, Cameron Parish, Louisiana; and (3) Smith Point, Chambers County, Texas. The study areas at each site consisted of a "disturbed" plot with reduced understory (i.e., structure, species composition, or both) primarily due to cattle grazing, paired with an adjacent "control" plot, where the understory had not been as affected. The fact that plots were adjacent is important, because adverse weather conditions or depleted energy reserves may restrict the movements, and thus the range of habitats available to migrants during stopover (Moore et al. 1993). Plots were 1.5 ha in size and were flagged at 25-m intervals to facilitate the running of strip transects and the collection of behavioral observations. A stratified random sampling scheme was used to identify differences in habitat structure between disturbed and control plots at each site. At each sampling location, we recorded the presence or absence of four habitat types within an imaginary cylinder with a diameter of 1 m: near-ground (0–0.5 m), understory (>0.5–2 m), subcanopy (>2–10 m), and canopy (>10 m). Control plots at Hackberry Ridge and Grand Chenier had a higher proportion of vegetation in the understory and subcanopy; proportion of canopy vegetation was similar between plots (Fig. 5). At Smith Point, no difference in understory structure occurred between plots; proportion of subcanopy vegetation was greater in the control plot (Fig. 5). Occurrence of grass and herbaceous cover was greater in the disturbed plot at Grand Chenier, greater in the control plot at Smith Point, and similar in the two plots at Hackberry Ridge (Fig. 5). Plant species diversity was reduced at all three disturbed plots (W. Barrow, unpubl. data).

To investigate how structural differences in

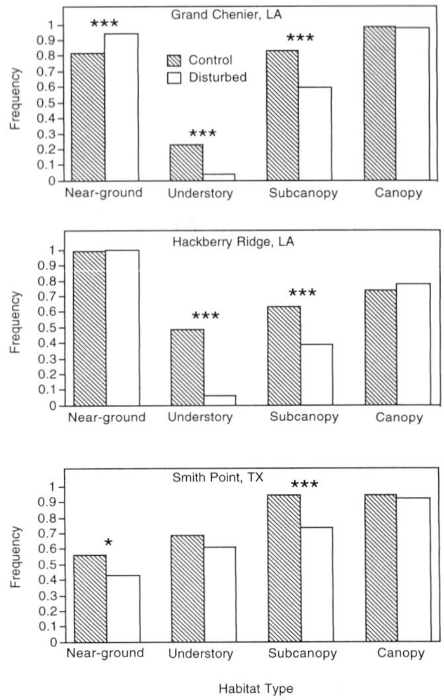

FIGURE 5. Comparison of proportion of habitat types between control and disturbed plots at three study sites. Significant differences of Z tests are indicated by asterisks (* = $P < 0.05$, *** = $P < 0.001$). Sample sizes in control and disturbed plots at Grand Chenier, Hackberry Ridge, and Smith Point are 213, 184; 336, 336; and 149, 165, respectively.

vegetation affect nearctic-neotropical migrants during stopover, migrants were first collectively examined and then separated into height guilds (ground, understory, and subcanopy/canopy) and substrate guilds (fruit/flower, air space, leaf litter, live foliage, suspended dead foliage, and bark). Height guild composition was determined by examining the height distributions of each species from the transect data sets in "control" plots. If at least 50% of the observations of a species were 0 m, 0–2.5 m, or >2.5 m from the ground, the species was placed in the ground, understory, or subcanopy/canopy guild, respectively (Appendix). Substrate guild composition was determined from observations on foraging migrants. Species predominantly foraging on one type of substrate (i.e., having at least 50% of their foraging observations on one substrate) were placed into their appropriate guild (Appendix). Classification into the "suspended dead foliage" guild required at least 50% of leaf-directed prey attacks to be on suspended dead leaves. Using strip transect data, Wilcoxon signed-rank tests were performed to compare the abundance of all migrants, early migrants, and

TABLE 1. COMPARISON OF MIGRANT USE OF CONTROL AND DISTURBED FOREST PLOTS (MEAN ± SD PER 1.5 HA) IN THE CHENIER PLAIN DURING SPRING 1993, 1994, AND 1995

Migrants	Grand Chenier		Hackberry Ridge		Smith Point	
	Control	Disturbed	Control	Disturbed	Control	Disturbed
1993	N = 74		N = 74		N = 69	
All migrants	9.7 ± 10.8*	6.5 ± 8.8	10.5 ± 12.4	18.1 ± 30.1	8.0 ± 10.0	10.5 ± 13.5*
Early migrants	4.4 ± 4.2*	2.9 ± 5.4	4.4 ± 4.7*	3.6 ± 5.4	7.6 ± 10.3	9.7 ± 14.5
1994	N = 67		N = 67		N = 75	
All migrants	9.5 ± 10.6*	4.2 ± 6.6	11.2 ± 14.8	13.1 ± 20.3	10.6 ± 13.8	16.4 ± 22.0*
Early migrants	2.4 ± 2.9*	0.4 ± 0.8	2.7 ± 2.6*	1.5 ± 2.7	4.2 ± 6.7	6.4 ± 9.8*
1995	N = 73		N = 73		N = 75	
All migrants	9.6 ± 14.2*	4.6 ± 8.1	11.1 ± 14.1	12.2 ± 18.3	3.0 ± 5.0	6.6 ± 10.6*
Early migrants	4.7 ± 8.4*	1.6 ± 3.7	6.7 ± 10.6*	4.3 ± 8.5	1.8 ± 3.7	4.5 ± 10.0*

* = P ≤ 0.05

substrate guilds in control versus disturbed plots at each site for spring migration 1993–1995. The high daily turnover rate of migrants using chenier forests (mean recapture rate = 7.6%; N = 5,146 initial captures in 1993) enabled differences in bird abundances between contrasting plots to be calculated on a daily basis. Significance was defined as $P < 0.05$ for all single tests of hypotheses. To avoid simultaneous inference from multiple tests of hypotheses, alpha levels were adjusted for each guild using the sequential Bonferroni method (Rice 1989, Beal and Khamis 1991). Family-wide alpha levels for the substrate guilds was thus $P < 0.008$.

Foraging studies were conducted at all sites to supplement abundance data. The frequencies with which birds used plant species on the Chenier Plain were obtained from quantified data on their foraging behavior. Plots were traversed systematically, and foraging maneuvers according to Remsen and Robinson (1990) were recorded. A variety of other habitat parameters, including the height, substrate, and plant species on which the last maneuver occurred, were also recorded. To ensure that foraging data were gathered from separate individuals, we did not collect data on two consecutive individuals of the same species and sex in a given 25-m^2 grid during a data gathering period (1–3 hrs per d).

Plant species use versus availability comparisons were made to determine which plant species were preferred by migrants. Vegetation was randomly sampled to obtain relative frequency and relative basal area coverage for each plant species at each site. We made these comparisons only in the control plots because they most closely resemble natural conditions; our interest here was to make recommendations for restoration of chenier forests.

ABUNDANCE OF MIGRANTS IN DEGRADED VERSUS INTACT HABITAT

Two of the questions this study seeks to answer that relate to restoration planning are: (1) what does the impact of understory degradation have on the abundance of trans-Gulf migrant landbirds using chenier forests, and (2) which plant species and structural features of chenier forests are preferred by migrants during stopover.

All migrants

No consistent trends were detected across sites for all migrants (Table 1). All migrants were more abundant in control plots each year at Grand Chenier and in disturbed plots each year at Smith Point (Table 1). Grand Chenier control plots had greater vegetation in the understory and less grass/herbaceous ground cover compared to Smith Point. The reduced vegetation near the ground may have afforded better foraging opportunities for those species that search for prey living in the leaf litter, whereas greater understory vegetation provided perch sites for migrants that forage within this stratum. Factors determining the greater use of the disturbed plots at Smith Point remain unclear. Understory vegetation structure was similar between plots; however, the species composition was different: cherry laurel dominated the control plot, and yaupon and *Rubus* sp. were dominant understory plants in the disturbed plot. Because of secondary compounds present in the foliage of cherry laurel, this plant harbors few leaf-chewing insects (W. Barrow, unpubl. data). This may account for the greater use of the disturbed plots at Smith Point when all migrants were combined.

Early migrants

When transect data from all sites were combined, distributions of migrants shifted from control to disturbed plots during the migratory season (Fig. 2). Early migrants, individuals detected before the second week of April (primarily Ruby-throated Hummingbird, White-eyed Vireo, Yellow-throated Vireo, Northern Parula, Yellow-throated Warbler, Black-and-white Warbler, Louisiana Waterthrush, and Hooded Warbler), were significantly more abundant in control plots at the two sites having significantly more understory vegetation in control plots (Fig. 2; Table 1). Food may be especially limiting during early spring and late fall migration (Martin and Karr 1990), suggesting that undisturbed habitats may be critical for early migrating species. Disturbed plots were predominantly used when migrants reached their greatest densities (Fig. 2). High concentrations of energy-depleted birds may cause competition for food resources at stopover sites (Moore and Yong 1991). Competition would explain the increased densities of birds in disturbed plots during the peak of migration, but not the greater use of disturbed habitats than intact habitats, as occurred in 1993. The cause of this apparent switch in proportional use of disturbed versus control plots during spring migration is unclear, but it may be due to reduced bird detectability in control plots after leaf-out; to the late influx of ground foraging birds (e.g., thrushes), which may prefer more open habitats (see below), or to short-term changes in environmental conditions (e.g., food resources). We feel the latter may play an important role at our study sites. If food became more abundant in the disturbed (vs. control) plots during the late migration period, then the change in bird abundance can be explained fairly well. Disturbed plots at two of our sites may provide more food resources during the time that migrant numbers were greater in the disturbed plots. At Hackberry Ridge, the disturbed plot had a higher density of hackberry trees than the control plot; at Grand Chenier, the disturbed plot had a greater density of honeylocusts than control plots (W. Barrow, unpubl. data). Both tree species can provide an extraordinary abundance of food resources from middle April through early May. An annual outbreak of lepidopteran (Geometridae) larvae occurs on Hackberry Ridge and is specific to hackberry trees (Moore and Yong 1991; W. Barrow, pers. obs.). The flowers of honeylocust are used by several nectarivorous species; the foliage apparently harbors an abundance of insects (e.g., honeylocust pod gall midge larvae [*Dasineura gleditchiae*]; W. Barrow, unpubl. data), as it is a preferred plant by feeding migrants (see below). The relations between food resource phenology, abundance, and habitat switching by *en route* migrants at our study sites provide an excellent opportunity for further research.

Foraging substrate guilds

Species that are specialists on certain foraging substrates are affected by alterations to vegetation of stopover habitats. Abundance of nectarivorous and frugivorous migrants was significantly greater in control plots versus disturbed plots at Grand Chenier (1993 and 1994) and Hackberry Ridge (all years; Table 2), the two sites with a reduced understory (Fig. 5). Grazing has been shown to negatively affect frugivores and nectarivores in temperate regions (Bock et al. 1984, Taylor 1986, Knopf et al. 1988) as well as tropical areas (Saab and Petit 1992). Because very few birds breed on cheniers (Moore and Yong 1991), nectarivorous and frugivorous migrants may play key roles in pollination and seed dispersal (Herrera 1984, Saab and Petit 1992). Species foraging on suspended dead leaves also were more abundant in control plots at Grand Chenier (all years) and Smith Point (1993 and 1994; Table 2). This would be expected because the denser understory and subcanopy vegetation in control plots would be more likely to trap dead leaves as they fall from the canopy. Ground foraging migrants that feed on insects and arthropods of the forest's leaf litter were more abundant on the control plot at Grand Chenier all years (Table 2). The reduced amount of grasses and herbs near the ground was apparently more suitable for those species that rummage through the leaf litter on the forest floor at this site. No consistent significant trends were observed across sites for the remaining substrate guilds (Table 2). Canopy vegetation remaining in the disturbed sites apparently provided sufficient foraging substrates for bark and live-foliage gleaners.

Bird Movements Within and Between Habitats

We can use measures other than bird density to evaluate habitat quality during the nonbreeding season (Winker et al. 1995). Based on observations of arriving trans-Gulf migrants in coastal woodlots, Moore et al. (1993) suggested that migrants "rank" alternative habitats during an initial exploratory phase shortly after arrival. Our results support this hypothesis. Nets were randomly placed in 25 m × 25 m blocks throughout the study sites. Number, arrangement, size, height and orientation of the nets were the same in control and disturbed plots and remained the same among years. Each day, we

TABLE 2. COMPARISON OF MIGRANT USE IN CONTROL AND DISTURBED FOREST PLOTS (MEAN ± SD PER 1.5 HA) IN THE CHENIER PLAIN DURING SPRING 1993, 1994, AND 1995

Substrate guilds	Grand Chenier Control	Grand Chenier Disturbed	Hackberry Ridge Control	Hackberry Ridge Disturbed	Smith Point Control	Smith Point Disturbed
1993	N = 74		N = 74		N = 69	
Leaf litter	2.2 ± 3.4*	1.8 ± 3.7	1.2 ± 2.7	4.4 ± 10.4*	1.4 ± 2.8	2.6 ± 4.4
Dead foliage	0.5 ± 1.0*	0.2 ± 0.5	0.2 ± 0.3	0.2 ± 0.4	0.4 ± 0.9*	0.3 ± 0.7
Bark	0.2 ± 0.5	0.2 ± 0.6	0.3 ± 0.5	0.4 ± 1.1	0.4 ± 0.8	1.0 ± 2.0
Live foliage	1.1 ± 2.0	1.3 ± 2.8	0.6 ± 1.4	1.5 ± 3.3	0.5 ± 1.2	0.9 ± 2.3
Fruit/flower	0.8 ± 1.8*	0.2 ± 0.8	3.1 ± 6.1*	1.4 ± 3.0	0.2 ± 0.5	0.1 ± 0.4
Air space	0.1 ± 0.3	0.3 ± 0.7	0.2 ± 0.8	0.4 ± 1.3*	0.1 ± 0.2	0.2 ± 0.9
1994	N = 67		N = 67		N = 75	
Leaf litter	1.9 ± 2.7*	1.0 ± 2.4	1.6 ± 2.9	2.4 ± 4.7	1.2 ± 2.3	2.7 ± 6.2
Dead foliage	0.6 ± 1.2*	0.0 ± 0.2	0.3 ± 0.7	0.3 ± 0.8	1.2 ± 2.5*	0.8 ± 2.2
Bark	0.3 ± 0.6*	0.1 ± 0.2	0.2 ± 0.5	0.3 ± 0.6	0.8 ± 1.5	1.0 ± 1.8
Live foliage	1.8 ± 3.3*	0.7 ± 1.7	0.9 ± 2.3	1.5 ± 3.3	0.7 ± 2.0	1.6 ± 3.6
Fruit/flower	0.9 ± 2.0*	0.2 ± 0.6	2.5 ± 4.1*	0.7 ± 1.6	1.0 ± 2.1	1.6 ± 4.0
Air space	0.1 ± 0.3	0.3 ± 0.6	0.2 ± 0.4	0.3 ± 0.5	0.1 ± 0.4	0.7 ± 1.3*
1995	N = 73		N = 73		N = 74	
Leaf litter	1.3 ± 2.3*	0.3 ± 0.9	1.5 ± 2.8	1.5 ± 3.0	0.5 ± 1.2	1.5 ± 2.9*
Dead foliage	0.4 ± 0.7*	0.1 ± 0.3	0.3 ± 0.7	0.2 ± 0.4	0.1 ± 0.3	0.1 ± 0.4
Bark	0.4 ± 0.7*	0.1 ± 0.4	0.3 ± 0.5	0.3 ± 0.7	0.2 ± 0.4	0.7 ± 1.6
Live foliage	1.9 ± 4.0	1.2 ± 2.8	1.2 ± 2.8	1.5 ± 2.9	0.4 ± 1.4	0.7 ± 1.7
Fruit/flower	1.1 ± 2.3	0.9 ± 2.0	2.2 ± 4.0*	1.3 ± 3.7	0.2 ± 0.8	0.2 ± 0.7
Air space	0.3 ± 0.7	0.3 ± 0.7	0.2 ± 0.6	0.3 ± 0.7	0.1 ± 0.5	0.4 ± 0.9

* = $P \leq 0.008$.

attempted to have the same net exposure in both control and disturbed plots. Capture data were standardized to captures per 500 net hours. At Smith Point, 63% of birds known to have moved between plots (based on recaptures and observations of color-marked individuals) switched from the disturbed to control plot (one-tailed binomial test, P = 0.01, N = 87). Worm-eating Warblers, specialists on suspended dead leaves, and Hooded Warblers, subcanopy foragers, frequently moved from disturbed plots to control plots (75%, P = 0.07, N = 12; and 64%, P = 0.21, N = 14, respectively). In contrast, Black-and-white Warblers, bark specialists, moved between plot types with equal frequency (47%, P = 0.81, N = 17). At Grand Chenier, 67% of migrants in 1993 (only year with sufficient sample size) known to have moved between plots settled in the control plot (one-tailed binomial test, P = 0.01, N = 31). These data are further evidence that birds use the disturbed and control plots differently. The Hackberry Ridge site was not included in this analysis because the plots along this ridge were separated by nearly 1 km.

The length of time that migrants stay in an area is another method to evaluate habitat quality. A comparison of recapture rates between contrasting habitats is a way to measure this (Winker et al. 1995). The paired plots situated along Hackberry Ridge had the greatest contrast in vegetative structure between disturbed and control plots. Woody vegetation at the Hackberry Ridge disturbed plot is essentially nonexistent below the canopy layer (primarily hackberry trees). Several subcanopy-height trees are present (mostly honeylocusts), leaf litter is reduced, and grass/herbaceous cover is abundant. The structure is not unlike an urban "park." At Hackberry Ridge, the proportion of birds recaptured on the control plot was higher than the disturbed plot for ground guilds in 1993 and 1994 and subcanopy/canopy guilds for all years; recapture rates were similar between the plots for understory birds (Fig. 6). Apparently, most species in our understory guild were able to shift their foraging activity down to the ground (e.g., Gray Catbird, Kentucky Warbler, Common Yellowthroat, and Hooded Warbler) or up to the canopy (e.g., Canada Warbler and Yellow-breasted Chat); others may have moved to more suitable habitat (e.g., Worm-eating Warbler; Chen 1996). No differences in recapture rates were found between contrasting plots at the other two study sites; this is not very surprising because the sample sizes were low and the habitat differences were much less than at Hackberry Ridge.

Future research on the suitability of chenier forests as stopover sites should focus on rates of

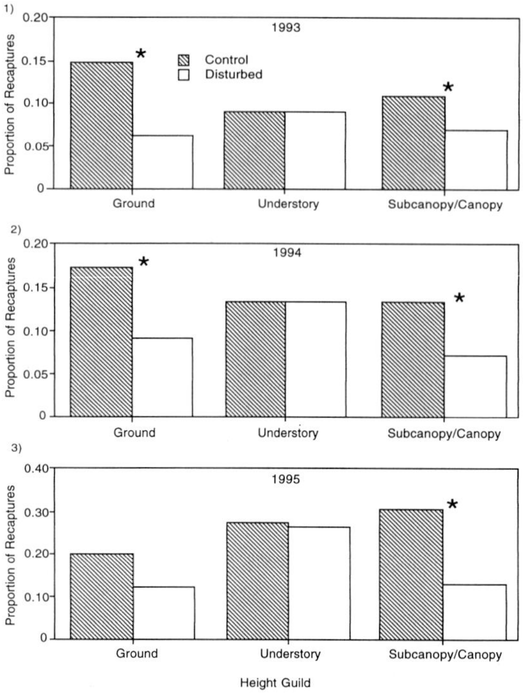

FIGURE 6. Comparison of proportion of recaptures per 500 net hours between control and disturbed plots at Hackberry Ridge. Significant differences of G-tests are indicated by asterisks (* = $P < 0.05$). Sample sizes of initial captures in control and disturbed plots for ground, understory, and subcanopy/canopy guilds for (1) 1993: 290, 315; 301, 147; and 785, 402; (2) 1994: 285, 243; 173, 87; and 662, 421; (3) 1995: 104, 78; 100, 27; and 215, 135, respectively.

weight gain for *en route* migrants during their stay in these contrasting vegetation conditions.

Important Food Plants

Land managers interested in restoring or rehabilitating habitat for *en route* nearctic-neotropical migrants need to know which plant species are important for these birds. From observations taken on foraging migrants, plant species use versus availability comparisons were made to determine which plant species would be most beneficial for *en route* nearctic-neotropical migrants.

Chi-square goodness-of-fit tests, performed on all plant species constituting more than 5% of the available vegetation at each site, indicated that the use of plant species by migrants differed significantly from the availability of plant species at each site ($P < 0.001$ for all sites). Following Neu et al. (1974), Bonferroni confidence intervals were calculated to determine significant preference or avoidance of individual plant species by migrants. All foraging analyses are from 1993 data only.

Understory vegetation

Shrubs common enough (i.e., constituted at least 5% of total plants on the control plots) to test statistically included yaupon (Smith Point) and deciduous holly (Grand Chenier), which were used in equal proportion to their availability by all migrants (Table 3). Yaupon was significantly preferred by understory species at Smith Point ($P < 0.05$, $N = 92$). Green hawthorn (Grand Chenier) was the only small tree used significantly more than available (Table 3). Cherry laurel (Smith Point) and Chinese tallow (Smith Point) were both significantly avoided by all migrants (Table 3). Understory species used cherry laurel in equal proportions to its availability ($P > 0.05$), and significantly avoided Chinese tallow ($P < 0.05$). Preliminary results from a study on arthropod-plant relations indicate that densities of arthropods are much lower in cherry laurel and Chinese tallow than in live oak, hackberry, and yaupon (W. Barrow and T. Spengler, unpubl. data). Cherry laurel is known to have a secondary chemical compound, prussic acid, in its foliage to deter herbivory (Vines 1960). Not surprisingly, 30% of the foraging observations occurred on the bark rather than the foliage of this tree. The mechanism deterring insects from associating with Chinese tallow is unknown, but apparently effective.

Canopy vegetation

Hackberry was used in larger proportion than any other plant species at both sites in which it occurred, and was significantly preferred at Grand Chenier (Table 3). However, hackberry was significantly avoided by understory species at Hackberry Ridge ($P < 0.05$, $N = 50$). Live oak was used in proportion to its availability at two sites, but was significantly avoided by all migrants (Table 3) at Grand Chenier and by understory species at Smith Point ($P < 0.05$, $N = 92$). The avoidance of both hackberry and live oak by understory species may be due to the lack of regeneration of these species. Because few live oak and hackberry saplings occur on these cheniers, foliage of these species are available only in the subcanopy/canopy stratum of the forest. The distribution of plant species on cheniers may also affect their use by foraging birds. When a habitat is primarily comprised of a single species, such as hackberry at Hackberry Ridge or live oak at Smith Point (Table 3), the benefits of foraging on the less abundant plant species may be outweighed by the costs associated with searching for uncommon or rare species. When plants are more evenly distributed

TABLE 3. Simultaneous Confidence Intervals for the Use of Plant Species by Nearctic-Neotropical Migrants at Three Control Plots in the Chenier Plain

	Expected proportion of use, P_{io}	Actual proportion of use, P_i	Bonferroni intervals for P_i
Grand Chenier		N = 442	
Deciduous holly	0.070	0.088	0.051 ≤ P ≤ 0.125
Green hawthorn	0.098	0.149	0.103 ≤ P ≤ 0.196*
Hackberry	0.151	0.253	0.197 ≤ P ≤ 0.310*
Honeylocust	0.062	0.075	0.040 ≤ P ≤ 0.109
Live oak	0.254	0.133	0.089 ≤ P ≤ 0.178*
Vine	0.092	0.145	0.098 ≤ P ≤ 0.192*
Other	0.178	0.085	0.049 ≤ P ≤ 0.123*
Hackberry Ridge		N = 303	
Hackberry	0.446	0.429	0.354 ≤ P ≤ 0.504
Honeylocust	0.020	0.109	0.062 ≤ P ≤ 0.156*
Live oak	0.057	0.053	0.019 ≤ P ≤ 0.087
Red mulberry	0.186	0.119	0.070 ≤ P ≤ 0.169*
Vine	0.173	0.211	0.149 ≤ P ≤ 0.273
Other	0.119	0.079	0.038 ≤ P ≤ 0.120
Smith Point		N = 274	
Cherry laurel	0.140	0.086	0.043 ≤ P ≤ 0.133*
Chinese tallow	0.101	0.040	0.009 ≤ P ≤ 0.071*
Live oak	0.464	0.460	0.380 ≤ P ≤ 0.539
Yaupon	0.263	0.325	0.250 ≤ P ≤ 0.400
Vine	0.013	0.066	0.026 ≤ P ≤ 0.105*
Other	0.017	0.022	0.000 ≤ P ≤ 0.045

N = Total number of foraging migrants recorded.
* $P < 0.05$.

throughout a habitat, such as at Grand Chenier (Table 3), avian foraging preferences may become more readily apparent.

Plants producing fruit or flowers during the spring migration season appear to be especially important to several migrant species. For instance, although red mulberry was significantly avoided at Hackberry Ridge when all species were combined (Table 3), some species fed almost exclusively on the fruit of this tree. Red mulberry fruit was eaten in 35 out of 50 independent foraging observations taken on Gray Catbirds and 14 of the 31 observations on Rose-breasted Grosbeaks. The flowers of the honeylocust tree appeared to be especially important to nectarivorous species, as well as many insectivorous species. Honeylocust was the only tree significantly preferred by migrants at Hackberry Ridge (Table 3). Although honeylocust only comprised 2% of the plant community at Hackberry Ridge, more than 48% of the foraging observations for Tennessee Warblers (N = 103), and more than 70% of the foraging observations for Baltimore Orioles (N = 18) and Orchard Orioles (N = 20) occurred on the flowers of this tree.

Vine tangles

Vines are an important habitat feature for migrants. Vines were preferred at all three sites, with a significant preference for vines by migrants occurring at two of the three sites (Table 4). When just vine species were compared to each other, Virginia creeper was the only vine that was used significantly less than expected by migrants (Table 4). The only vine significantly preferred to other vines was Japanese honeysuckle, an exotic species (Table 4). Although grape vine was not significantly preferred by all migrants, it was used more than any other vine species. Because vines easily entrap falling dead-leaf clumps, they are also important food patches for the Worm-eating Warbler and several species in the genus *Vermivora.* Vine tangles have been previously noted as an important foraging substrate in bottomland hardwood forests for at least some species of breeding migrants (Barrow 1990, Moser et al. 1990, Pashley and Barrow 1992), and appear to be important for *en route* migrants as well.

Exotic plants

All nonindigenous trees and shrubs such as chinaberry, Chinese tallow, cherokee rose, and lantana were used less than expected by migrants. At Smith Point, the only site in which an exotic was abundant enough to test statistically, Chinese tallow was used significantly less than expected by migrants (Table 3). Invasion of ex-

TABLE 4. SIMULTANEOUS CONFIDENCE INTERVALS FOR THE USE OF VINE SPECIES BY NEARCTIC-NEOTROPICAL MIGRANTS AT THREE CONTROL PLOTS IN THE CHENIER PLAIN

Vine species	Expected proportion of use, P_{io}	Actual proportion of use, P_i	Bonferroni confidence intervals for P_i
Grape vine	0.529	0.451	$0.356 \leq P \leq 0.545$
Greenbrier	0.049	0.052	$0.009 \leq P \leq 0.094$
Japanese honeysuckle	0.080	0.164	$0.094 \leq P \leq 0.235$*
Lady's eardrop	0.004	0.019	$0.000 \leq P \leq 0.045$
Poison ivy	0.016	0.047	$0.007 \leq P \leq 0.087$
Rattan vine	0.189	0.207	$0.129 \leq P \leq 0.284$
Trumpet creeper	0.042	0.056	$0.012 \leq P \leq 0.100$
Virginia creeper	0.089	0.005	$0.000 \leq P \leq 0.018$*

N = Total number of foraging migrants recorded.
* $P < 0.05$.

otic plants may pose one of the most serious threats to the integrity of chenier forests. Once permanently established, exotic organisms have only been successfully eradicated in a few expensive and labor-intensive instances (Coblentz 1990, Westman 1990). While some scientists argue that all exotics should be removed (Coblentz 1990), others suggest that the contribution of exotic species to wildlife communities should first be examined, and only those shown to severely degrade native habitats be removed (Westman 1990).

Due to their low abundance on our sites, cherokee rose, chinaberry, and lantana need further investigation before substantive results on migrant use of these exotic species can be obtained. We have only preliminarily examined exotic plant/avian/insect relationships, and that has been restricted to the spring migration season. Exotic plants (e.g., Chinese tallow), and native plants (i.e., cherry laurel and palmetto) that are avoided by spring migrants may provide important food sources (fruit pulp, waxy epicarp) or cover for birds using these habitats in fall and/ or winter.

SUMMARY FOR THE CHENIER PLAIN FIELD STUDY

Because the majority of forested habitats in the Chenier Plain were altered close to a century ago, the effects that this loss of stopover habitat had on neotropical landbird migrants cannot be determined. The conservation value of a chenier should thus not be compared to the original intact chenier forest type, but to the land presently surrounding them. Clearly, a disturbed forest, even one intermingled with exotic plants, is more desirable than pasture or cotton fields. As revealed in these analyses, most forest-dependent migratory birds are tolerant of at least some degradation of chenier forests during migration. However, these results show that subtle differences in vegetation composition and structure of the understory layer of these forests can result in differential use by some *en route* migrants. Because of species-specific habitat requirements, the effect of understory reduction is not similar for all nearctic-neotropical migrants, and each condition may provide benefits for at least some species.

Cheniers with a disturbed understory were less frequently used by early migrants, dead-leaf foragers, frugivores and nectarivores, but had little or no effect on most other nearctic-neotropical migrant species. However, our study may have underestimated actual differences in habitat use by migrants. Bird detectability was greater in the open, disturbed plots, and most cheniers in the region were more heavily grazed than the treatment plots used in this study. Thus, the results from this study are probably conservative in their estimation of the effects habitat degradation has on *en route* nearctic-neotropical migrants.

Hackberry, red mulberry, honeylocust, green hawthorn, vine tangles, and other plants that fruit or flower during the spring-migration period appear to be important microhabitat features for *en route* nearctic-neotropical migrants in the Chenier Plain. Our foraging data indicate that the structure and floristics of these forests may influence the foraging opportunities available to migrants and therefore affect how successfully they can exploit these stopover habitats. Because species may specialize on different foraging substrates, stopover habitats with diverse plant communities that are able to produce a variety of nectar, catkins, fruits, and seeds, should be best equipped to produce sufficient food resources for all migratory species (Moore and Simons 1992a, Moore et al. 1993).

The decision to rehabilitate a forested stopover site depends on several factors, such as funding levels and land ownership. Otherwise, the two most important considerations are: (1)

the relative status of forest conditions in the migration pathway or landscape under consideration, and (2) the relative status of birds that may be negatively impacted by habitat disruption in the landscape under consideration. Given that the understory structure and regeneration of chenier forests has been so greatly reduced, and that several species considered here to be sensitive to understory degradation are believed to have declining population trends (Robbins et al. 1989b), restoration and rehabilitation should be a priority.

RESTORATION RECOMMENDATIONS AND DIRECTIONS FOR FURTHER WORK

Ultimately, the goal is to rehabilitate *en route* habitat at key stopover locations in a manner that ensures their suitability to the birds that depend on these wooded sites for food, rest, and shelter each year during spring and fall migration periods. Because of species-specific microhabitat preferences, no single restoration or rehabilitation plan will have a similar effect on all nearctic-neotropical migrants, and each management practice will provide benefits for at least some species. However, because understory vegetation is the most degraded, restoration/rehabilitation projects should concentrate their efforts on this habitat component.

Recommendations of Specific Management Actions for the Chenier Plain

Rehabilitation of coastal woodlands will depend on local involvement. Private citizens own 95% of the cheniers of Louisiana and Texas. We recommend establishment of working relationships with the Cameron Parish, Louisiana, and Chambers County, Texas, Cattlemen's Associations to discuss opportunities to modify grazing schedules that will encourage vegetative regeneration while still meeting the economic needs of ranchers. It will be important for conservationists to keep in mind that if it were not for the cattle ranching industry of southwest Louisiana, the forest remnants that are so important today may otherwise be private homesteads or petrochemical plants.

On cheniers where landowners want to restore woodlands for migratory landbirds, we recommend the following:

1. On cheniers where little or no understory exists, or where little or no regeneration is occurring, grazing pressure should be reduced until vegetation recovers. Once a chenier recovers, grazing may be allowed under constraints, preferably low-density, winter-only grazing (November 1–March 31). If the lack of understory is a result of overbrowsing by deer, the deer population should be reduced.

2. Portions of each chenier, especially the remaining, relatively intact chenier forests, should be permanently protected from deer, cattle and goats. These reserves, if strategically dispersed, could serve as seed sources as well as provide structurally complex and diverse habitat for *en route* nearctic-neotropical migrants.

3. The suitability of recent, human-created habitats (e.g., tallow woodlands, levees, and spoil banks) in the Chenier Plain needs to be evaluated. Maintenance of the vegetation types on levees and spoil banks would not replicate historic conditions, but it would provide woody vegetative cover that may compensate for upland sites that can no longer be restored or rehabilitated.

4. On degraded cheniers, we recommend reseeding or planting trees such as live oak, hackberry, honeylocust, and red mulberry; shrubs such as yaupon, sweet acacia, deciduous holly, and green hawthorn; and vines such as poison ivy, rattan, trumpet creeper, grape, and greenbrier. Studies of habitat and plant species use during fall migration need to conducted before a complete list of beneficial plants can be compiled.

5. Although rehabilitation is recommended for the forested cheniers, it may not be as feasible for native grasslands (i.e., coastal prairie). Most of this habitat has been converted to agriculture, pasture, or is now dominated by tallow woodlands. Research focused on habitat relations of migrants requiring grassland/prairie habitat during migration is needed.

Recommendations for Restoration and Rehabilitation of Stopover Habitats for Nearctic-neotropical Migrants

1. Inventory all key stopover locations. Determine vegetation characteristics (i.e., plant community type) and level of disturbance.

2. Collaboration among restorationists, conservationists, researchers, and landowners is essential. Technology for restoration/rehabilitation of coastal forested habitats, including exotic plant species control, is in its infancy. Due to the considerable time, expense, and effort associated with restoration projects, we suggest that information and technological advancement from previous restoration projects be exchanged among agencies, organizations, and landowners.

3. Many questions concerning the habitat requirements of migratory birds during stopover, especially during fall, remain unanswered. A recent study by Leberg et al. (1996) suggests that water may be a limiting factor for *en route* migrants that have just crossed the Gulf. Further studies examining the importance of water sources in stopover habitats to *en route* migrants

are needed. In addition, the response of migratory birds to various restoration practices and exotic species, as well as the importance of staging areas near ecological barriers in Mexico and Central America (Parker 1994), still need to be determined.

4. Work with state and local mosquito abatement programs to develop a plan that would lessen the possibility of indirect impacts on the non-target arthropod/insect community inhabiting wooded habitats. Avoidance of aerial spraying immediately prior to and during migration is recommended.

5. Because nearctic-neotropical migrants are so diverse in their foraging strategies, structurally complex and diverse plant communities able to provide sources of both fruit and nectar during the migration seasons should be protected or established wherever possible. Encourage local communities situated in key stopover locations to landscape with native plants indigenous to the region.

6. Restoration and rehabilitation of stopover habitats depend on the development of partnerships with landowners of the region; their understanding and cooperation is essential.

In conclusion, restoration and rehabilitation of stopover habitat in the Chenier Plain, as well as at other key stopover locations, will require the introduction of new technology, insights, and most importantly, enlightened management techniques through cooperative efforts with the local people.

ACKNOWLEDGMENTS

We are grateful to the Stream family in Louisiana for permission to do research on their property. Permission to do research in Texas was granted by Texas Parks and Wildlife Department through T. Tourney, Candy Abshier WMA. Stream Property Management Co., Rockefeller Wildlife Refuge, Louisiana Department of Wildlife and Fisheries, and Tennessee Gas Co. provided invaluable assistance with this study. The fieldwork was supported by Houston Audubon Society and U.S. Fish and Wildlife Service. Bill Goulet photographed and provided the aerial images of the cheniers. This study would not have been possible without the capable help of our 23 field assistants. J. Lind and S. Lind assisted with all phases of the study, especially data entry and management. V. Moseley provided several obscure reports that were cited in the manuscript. We thank F. Moore for many insightful discussions of the topic. C. Cordes, G. Gomez, R. Greenberg, K. Gutzwiller, F. Moore, D. Pashley, D. Richard, B. Vairin, and an anonymous reviewer provided many useful comments on the manuscript. We are especially grateful to N. and S. Baughman, J. and W. Erbelding, G. Nunez and family, D. Richard, and T. Tourney for use of their facilities and help with logistics.

APPENDIX. SPECIES COMPOSITION AND TOTAL NUMBER OF MIGRANTS DETECTED ON TRANSECTS AT THREE STUDY SITES FOR SPRING MIGRATION SEASONS 1993–1995

	1993	1994	1995
Black-billed Cuckoo (*Coccyzus erythropthalmus*)[c]	8	2	0
Yellow-billed Cuckoo (*Coccyzus americanus*)[c]	105	194	161
Common Nighthawk (*Chordeiles minor*)[c]	1	1	2
Ruby-throated Hummingbird (*Archilochus colubris*)[c,d]	102	246	133
Eastern Wood-Pewee (*Contopus virens*)[c,e]	125	121	168
Acadian Flycatcher (*Empidonax virescens*)[c]	29	31	10
Great Crested Flycatcher (*Myiarchus crinitus*)[c]	23	45	19
Eastern Kingbird (*Tyrannus tyrannus*)[c,e]	46	72	48
White-eyed Vireo (*Vireo griseus*)[c]	347	265	253
Yellow-throated Vireo (*Vireo flavifrons*)[c]	31	23	70
Warbling Vireo (*Vireo gilvus*)[c]	4	0	3
Philadelphia Vireo (*Vireo philadelphicus*)[c]	15	14	23
Red-eyed Vireo (*Vireo olivaceus*)[c,g]	519	424	511
Blue-gray Gnatcatcher (*Polioptila caerulea*)[c]	216	222	120
Veery (*Catharus fuscescens*)[a,f]	84	174	91
Gray-cheeked Thrush (*Catharus minimus*)[a,f]	47	127	47
Swainson's Thrush (*Catharus ustulatus*)[a,f]	444	246	147
Wood Thrush (*Hylocichla mustelina*)[a,f]	774	501	234
Gray Catbird (*Dumetella carolinensis*)[b]	782	612	357
Blue-winged Warbler (*Vermivora pinus*)[c,h]	56	139	47
Golden-winged Warbler (*Vermivora chrysoptera*)[c,h]	14	17	12
Tennessee Warbler (*Vermivora peregrina*)[c,d]	277	291	341
Nashville Warbler (*Vermivora ruficapilla*)[c]	1	4	0
Northern Parula (*Parula americana*)[c]	231	143	84
Yellow Warbler (*Dendroica petechia*)[c]	22	22	55
Chestnut-sided Warbler (*Dendroica pensylvanica*)[c,g]	63	70	71
Magnolia Warbler (*Dendroica magnolia*)[c,g]	132	335	258
Cape May Warbler (*Dendroica tigrina*)[c]	2	3	2
Black-throated Green Warbler (*Dendroica virens*)[c]	32	87	44
Blackburnian Warbler (*Dendroica fusca*)[c]	34	16	17
Yellow-throated Warbler (*Dendroica dominica*)[c]	18	4	19
Bay-breasted Warbler (*Dendroica castanea*)[c]	45	219	121
Blackpoll Warbler (*Dendroica striata*)[c]	10	26	6
Cerulean Warbler (*Dendroica cerulea*)[c]	12	8	17
Black-and-white Warbler (*Mniotilta varia*)[c,i]	279	292	226
American Redstart (*Setophaga ruticilla*)[c]	92	181	150
Prothonotary Warbler (*Protonotaria citrea*)[c,g]	84	44	110
Worm-eating Warbler (*Helmitheros vermivorus*)[b,h]	152	224	102
Swainson's Warbler (*Limnothlypis swainsonii*)[a]	1	12	8
Ovenbird (*Seiurus aurocapillus*)[a,f]	246	291	209
Northern Waterthrush (*Seiurus noveboracensis*)[a,f]	130	119	66
Louisiana Waterthrush (*Seiurus motacilla*)[a,f]	45	6	66
Kentucky Warbler (*Oporornis formosus*)[b]	120	133	81
Mourning Warbler (*Oporornis philadelphia*)[c]	1	3	3
Common Yellowthroat (*Geothlypis trichas*)[b]	83	61	43
Hooded Warbler (*Wilsonia citrina*)[b]	464	428	296
Canada Warbler (*Wilsonia canadensis*)[b]	52	25	6
Yellow-breasted Chat (*Icteria virens*)[b]	13	3	15
Summer Tanager (*Piranga rubra*)[c]	79	63	150
Scarlet Tanager (*Piranga olivacea*)[c]	65	116	87
Rose-breasted Grosbeak (*Pheucticus ludovicianus*)[c,d]	86	111	134
Blue Grosbeak (*Guiraca caerulea*)[c]	10	13	73
Indigo Bunting (*Passerina cyanea*)[a]	961	677	644
Painted Bunting (*Passerina ciris*)[c]	49	28	25
Orchard Oriole (*Icterus spurius*)[c,d]	301	119	157
Baltimore Oriole (*Icterus galbula*)[c,d]	77	57	92

[a] Ground Guild.
[b] Understory Guild.
[c] Subcanopy/Canopy Guild.
[d] Fruit/Flower Foraging Guild.
[e] Air Space Foraging Guild.
[f] Leaf Litter Foraging Guild.
[g] Live Foliage Foraging Guild.
[h] Suspended Dead Leaf Foraging Guild.
[i] Bark Foraging Guild.

Studies in Avian Biology No. 20:88–98, 2000.

LANDBIRD MIGRATION IN RIPARIAN HABITATS OF THE MIDDLE RIO GRANDE: A CASE STUDY

DEBORAH M. FINCH AND WANG YONG

Abstract. Growing human populations and rapid ecological changes threaten the sustainability of the middle Rio Grande, a river corridor important to numerous species of wintering, breeding, and migrating waterfowl, shorebirds, and songbirds. We review the vegetational and human history of the middle Rio Grande, substantiate the importance of this system to landbirds in migration, relate patterns and trends of migratory bird populations to variation of stopover habitats, and present new data on stopover habitat use and biology of landbird migrants. We supplemented our review of the literature by evaluating banding and survey data from a new study we implemented on the river in 1994, and we re-analyzed population trend data from unpublished banding records. Analyses of data from Hink and Ohmart (1984) and our own study showed that assemblages of migratory landbirds varied in species richness and abundance among seasons and among cottonwood-willow, mesquite, salt cedar, Russian olive, drainage, and agricultural habitats. Our fat deposition data demonstrated that migrating landbirds use the bosque to replenish energy stores during travel. Stopover along the middle Rio Grande may be especially important for those species that migrate across the Chihuahuan Desert. We suggest that spatial and temporal changes in habitat cover, structure, and composition of the middle Rio Grande bosque have potential to influence habitat use, food availability, health and survival during migration, and ultimately, success of future populations of stopover migrants.

Key Words: habitat use, Middle Rio Grande, migratory landbirds, mist-netting, riparian habitats, stopover.

Hot and cold deserts, grasslands, and shrubsteppe dominate much of the interior midwestern and western United States, forming the Great Plains, the Great Basin, and the Chihuahuan, Mohave, and Sonoran Deserts (Allen 1967, Bender 1982, Brown 1985). Climate, water, and people are primary forces driving ecological systems in aridland environments of the West, influencing the amounts, varieties, patterns, and persistence of plant communities (Finch and Tainter 1995a). These lowland environments are inhospitable to many animal species that require habitats having greater amounts of moisture, forage, and structure, affecting the abundance and richness of resident animal communities.

Birds that travel for long distances over these arid deserts and plains frequently follow the rivers and streams that dissect these landscapes, stopping over in riparian habitats that provide water, food, and cover from sun and predators (Wauer 1977). The Bosque del Apache Wildlife Refuge in the middle Rio Grande Valley is an example of a famous stopover site for migrating and wintering cranes and waterfowl. Less well-understood are the habitat use patterns and values of western riparian corridors for migrating songbirds (Stevens et al. 1977). In contrast, several studies have documented the high species richness and abundance of breeding songbirds in desert and plains riparian woodlands (Knopf et al. 1988, Hodorff et al. 1988, Finch 1989, Finch and Ruggiero 1993, Rosenberg et al. 1991). Only recently has this kind of information been effective in guiding policy development for the conservation and restoration of riparian habitats (e.g., Bosque Biological Management Plan for the Middle Rio Grande; Crawford et al. 1993). In the Southwest in particular, where water is scarce, drought is common, livestock grazing is pervasive, and human populations are growing rapidly, desert river systems are swiftly changing beyond recognition (Dick-Peddie 1993, Scurlock 1995). The recent advent of restoration and recovery plans for southwestern riparian ecosystems appears driven by the federal listing of threatened and endangered species. It is unfortunately a sign of our times that a species such as the Bell's Vireo (*Vireo bellii*) or the Willow Flycatcher (*Empidonax traillii*) must first become endangered in part or all of its range before the deterioration of an entire ecosystem is addressed.

DeSante and George (1994) concluded that the most important factor contributing to the population declines of western landbird species over the last 100 years was loss or destruction of riparian and marsh habitats, affecting as many as 20 western species, or 26% of those western species whose numbers have declined. By abstracting those landbird species that use riparian habitats to migrate and breed from DeSante's and George's overall list of declines of species state-by-state, we distinguished a consistent pattern among species in where they experienced population decreases. Specifically, declines of western bird species were clustered in California, Arizona, and other southwestern states (Table 1). Likewise, a recent report by Flather and

TABLE 1. STATES WITH POPULATION DECREASES OVER THE LAST 100 YEARS OF LANDBIRD SPECIES THAT MIGRATE AND BREED IN RIPARIAN ZONES OF THE WESTERN UNITED STATES (MODIFIED FROM DESANTE AND GEORGE 1994)

Species	Scientific Name	State
Yellow-Billed Cuckoo (N)[a]	*Coccyzus americanus*	*BC WA OR* **CA NV ID** UT AZ[b]
Willow Flycatcher (N)	*Empidonax traillii*	**CA AZ NM**
Vermillion Flycatcher (T)	*Pyrocephalus rubinus*	**CA NV**
Bank Swallow (N)	*Riparia riparia*	CA
Bell's Vireo (N)	*Vireo bellii*	**CA AZ**
Lucy's Warbler (N)	*Vermivora luciae*	CA AZ
Yellow Warbler (N)	*Dendroica petechia*	OR **CA AZ**
Common Yellowthroat (N)	*Geothlypis trichas*	CA AZ
Yellow-breasted Chat (N)	*Icteria virens*	CA **NV**
Summer Tanager (N)	*Piranga rubra*	CA AZ
Song Sparrow (T)	*Melospiza melodia*	AZ

[a] N = Nearctic-neotropical migrant, T = Temperate migrant. (Partners in Flight 1992).
[b] *Italic* = Extirpated, **Bold** ≥ 50% population decline.

Joyce (1994) on endangerment patterns of plant and vertebrate species graphically showed that the Southwest and California had higher concentrations of federally-listed threatened and endangered bird species than most other regions of the United States. We suggest that rapid environmental changes, coupled with high levels of endemism, may explain high endangerment rates in the Southwest.

We elected to focus our paper on southwestern riparian woodlands because nowhere has the decline of environmental quality been more evident. We highlight the middle Rio Grande, a river system considered by many to be in jeopardy (American Rivers 1993). In this paper, we describe the vegetational and human history of the middle Rio Grande system, review patterns and trends of migratory bird populations in relation to avian habitat use and vegetation change, present new data on stopover biology of migrants, and identify management problems. Our primary intent was to summarize the literature, but so little information was available on migrating birds along the Rio Grande or in other southwestern systems that we decided to supplement the review by evaluating banding and survey data from a new study we implemented on the river in 1994. We also re-analyzed population trend data from an unpublished study. Thus, our presentation covers a mixture of existing literature and new data, the latter of which we plan to publish in more detail elsewhere (e.g., Yong and Finch 1997b).

HISTORY OF CHANGE

The Rio Grande, one of two major river systems that drain the Southwest, originates in the mountains of Colorado and runs south about 3,220 km into the Mexico states of Chihuahua and Coahuila. The river and its associated riparian vegetation meander through an arid landscape of desert grasslands, canyons, arroyos, and mesas. It links natural environments from the highlands of the Rocky Mountains to the lowlands of the Chihuahuan Desert, and serves as an important dispersal route or corridor for many plants and animals.

The middle Rio Grande valley extends approximately 260 km from Cochiti Dam to Elephant Butte Reservoir, New Mexico (Finch and Tainter 1995b; Fig. 1); its width ranges from 1.5 km to about 10 km (Crawford et al. 1993). From north to south, it traverses three major biotic communities: Great Basin Grassland, Semidesert Grassland, and Chihuahuan Desertscrub. Historically, the middle Rio Grande contained one of the most extensive riparian gallery forests of cottonwoods (*Populus deltoides, P. fremontii*) in the southwestern United States (Howe and Knopf 1991). Associated with cottonwoods were a variety of native floodplain-adapted trees and shrubs including Goodding willow (*Salix gooddingii*), coyote willow (*Salix exigua*), New Mexico olive (*Forestiera neomexicana*), seepwillow (*Baccharis glutinosa*), wolfberry (*Lycium torreyi*), indigo bush (*Amorpha fruticosa*), and screwbean mesquite (*Prosopis pubescens*). The understory included grasses such as *Bouteloua* spp., *Sporobolus* spp., rabbitbrush (*Chrysothamnus nauseosus*), tumbleweed (*Salsola kali*), snakeweed (*Gutierrezia microcephala*), arrowweed (*Tessaria sericea*), sagebrush (*Artemisia* spp.), *Kochia* spp., and other forbs.

Human populations have increased dramatically along the middle Rio Grande since European settlement (Scurlock 1995, 1998). The floodplain riparian vegetation along the river has been impacted more by human activities than any other vegetation type in New Mexico (Dick-Peddie 1993). Current middle Rio Grande flood-

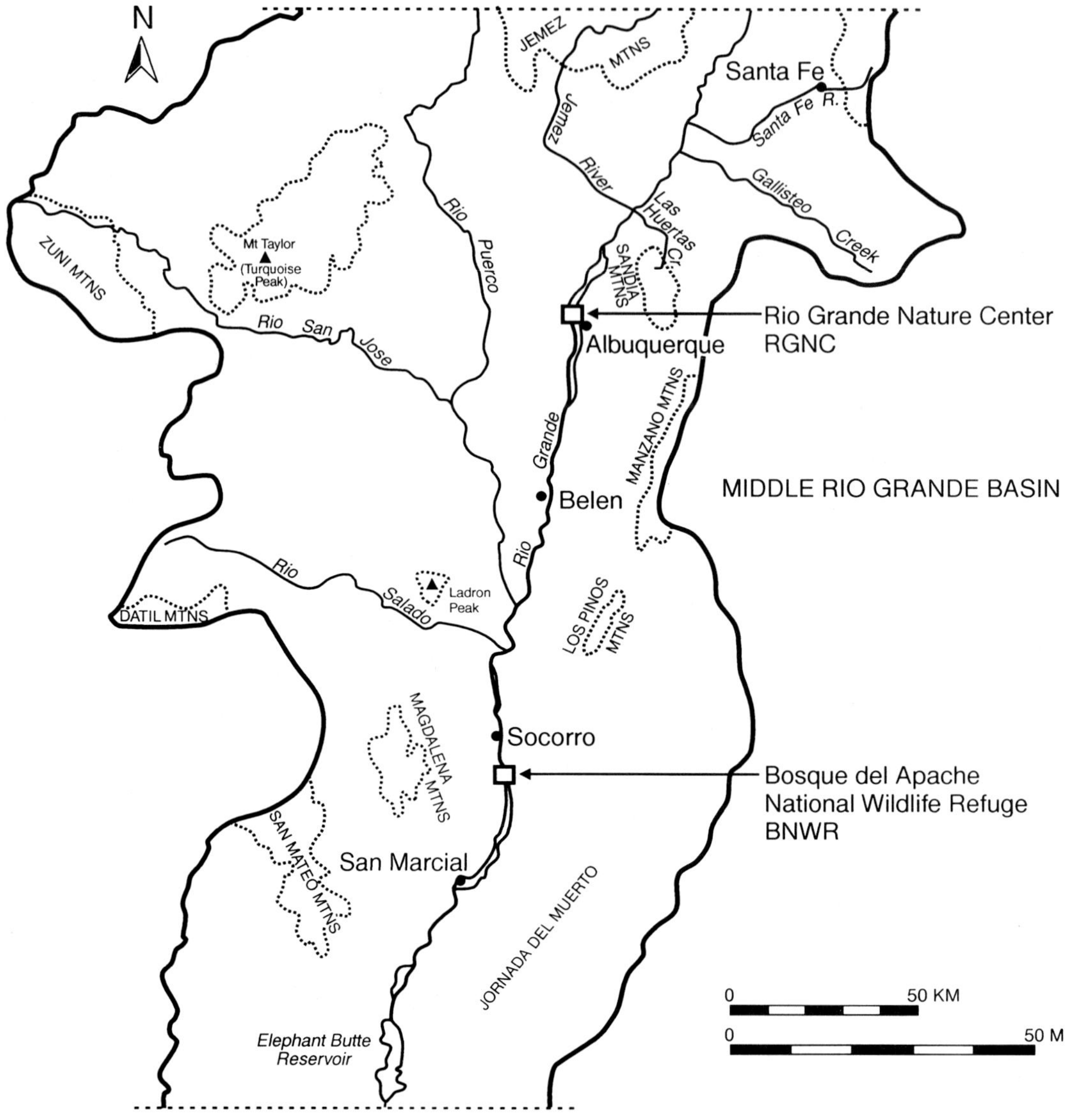

FIGURE 1. Map of the middle Rio Grande valley with square symbols marking locations of two banding stations, Rio Grande Nature Center and Bosque del Apache National Wildlife Refuge. Adapted from Scurlock (1998).

plain vegetation greatly differs in both composition and extent from pre-European plant communities owing to human-induced hydrological and ecological changes during the last two centuries (Bullard and Wells 1992; Scurlock 1995, 1998). While livestock grazing and timber and firewood harvesting reduced some of the existing woods and understory vegetation, the construction of riverside drains, levees, and irrigation structures in the early 1900s lowered the water table, draining much of the riverside wetland habitat and allowing further agricultural development of the floodplain (Wozniak 1995, 1998). Channelization and dam construction controlled annual floods and suppressed the regeneration of flood-dependent native species, particularly cottonwoods. Although cottonwood and willow were, and remain, dominant vegetation, they have been reduced to a narrow band of mid- to old-age forest stands between levees in the middle Rio Grande floodplain. Introduction and escape of saltcedar (*Tamarix chinensis*) and Russian olive (*Elaeagnus angustifolia*) since the turn of the century have further changed the

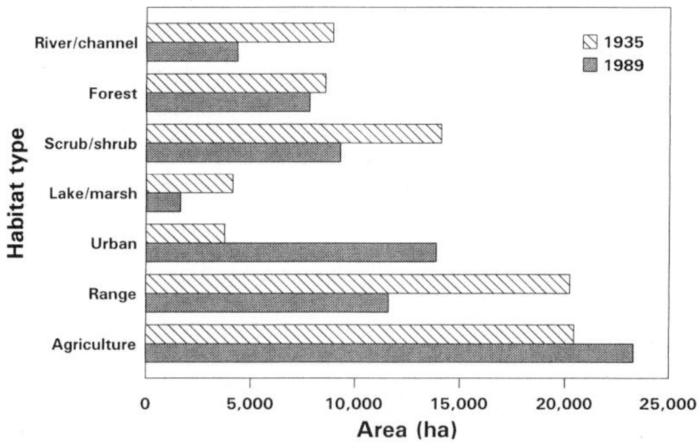

FIGURE 2. Changes in habitat availability along the middle Rio Grande between 1935–1989. Data were summarized from survey maps of the National Wetlands Inventory, Fish and Wildlife Service, U. S. Department of the Interior (Crawford et al. 1993).

composition and dominance of woody species, and the structure and successional stages of plant communities along the middle Rio Grande (Hunter et al. 1988, Dick-Peddie 1993).

More recently, growing human populations and associated housing and development have intruded into previously unoccupied and undisturbed areas, causing further modification to the riparian woodland, referred to locally as "the bosque" (Spanish for forest). Although the middle Rio Grande bosque appears continuous from the air, it is interspersed with residential areas, recreational parks, powerlines, bridges, road and trail networks, dams and diversion structures, exotic woody plants, croplands, pastures, and protected wildlife refuges (Finch et al. 1995). Maps comparing vegetation cover along the middle Rio Grande in 1935 to that in 1989 (National Wetlands Inventory maps prepared for Crawford et. al 1993), documented substantial changes in all cover types of the five river reaches (Cochiti, Albuquerque, Belen, Socorro, and Bosque del Apache) surveyed. When we totaled and compared cover values in 1935 and 1989 for each habitat across all five reaches, habitat turnovers over the 54-yr period were evident: river channel (−106% of 1935 channel), forest (−8%), scrubland (−50%), "range" grassland (−75%), lake/marsh (−150%), urban cover (+270%) and agriculture (+14%) (Fig. 2). The low amount of change in forest cover reported by the National Wetlands Inventory (Crawford et al. 1993) is somewhat misleading because the inventory did not distinguish native woodland from areas having introduced saltcedar or Russian olive, species that can exist as dominant plant communities or as understory layers in cottonwood communities. If the increased covers of introduced plant species were accounted for, the

TABLE 2. PREDICTED HUMAN POPULATION CHANGES IN THE MIDDLE RIO GRANDE VALLEY (MODIFIED FROM CRAWFORD ET AL. 1993)

County	1991	2020	Change	% change
Sandoval	63,841	128,996	65,155	+102
Bernalillo	481,689	614,265	132,576	+27
Valencia	45,545	91,831	46,286	+102
Socorro	14,804	21,216	6,412	+43
Total	605,879	856,308	250,429	+41

actual percent change in native forests would be much higher (e.g., Mount et al. 1996).

In a national report on endangered ecosystems of the United States, Noss et al. (1995) concluded that riparian forests in New Mexico were endangered. The environmental organization American Rivers (1993) identified the Rio Grande as one of the "most endangered rivers in North America". Alteration of river and riparian resources in the Southwest is associated with human population growth and accompanying land usages and impacts (Scurlock 1995 1998; Wozniak 1995, 1998). Human populations are predicted to almost double in size in the middle Rio Grande Basin by the Year 2020 (Table 2). The actual numbers of people may seem insignificant compared to large metropolitan areas such as Los Angeles or New York City, but given additional factors such as climate change in this arid region, declining water supply of the Albuquerque aquifer, deteriorating quality of the water and land, and geomorphological and land ownership constraints to urban growth, many scientists and land managers question whether ecosystems, human populations, and biological diversity in the Basin can be sustained (Finch and Tainter 1995a). Because migrating birds contribute in a vital way to the rich biological diversity of southwestern riparian systems (Rosenberg et al. 1991, Yong et al. 1995), no assessment of the sustainability of the middle Rio Grande valley would be complete without an account of this dynamic biological resource (Finch et al. 1995).

HABITAT VALUE TO MIGRATING BIRDS

River corridors may be more important as travel pathways for transient migrants in arid regions than waterways in areas with greater moisture and vegetation (Wauer 1977) because aridland river habitats offer fueling resources and shelter to birds from weather and predators. Although the dynamics of landbird migration along the Rio Grande have never been quantified, it is clear from the frequent displays and guided tours at local visitor centers that most refuge managers, park officials, and birders are fully aware that the middle Rio Grande provides recreational opportunities for observing numerous species of waterfowl, shorebird, and songbird in migration. Terrestrial riparian habitats along the Rio Grande provide diverse stopover sites for migratory landbirds that use the Great Plains-Rocky Mountain flight route (Finch et al. 1995). Yong et al. (1995) postulate that the hot and dry conditions of the Chihuahuan desert in Mexico and New Mexico may cause birds to funnel through the Rio Grande channel in search of available food, water, cover, and suitable north-south routing (see also Wauer 1977).

A review of earlier studies identified at least 241 landbird species that use the middle Rio Grande valley (Finch et al. 1995). Only 54 (23%) of these species are residents, and the rest are nearctic-neotropical and short distance migrants. Migrants include breeding residents (54 species, 22%) that are present in late spring and summer; winter residents (52 species, 22%) that are present for varying lengths of time between September and April; and transient species (71 species, 30%) that migrate through the valley in large numbers during spring and fall. Using categories of migratory status defined by Partners in Flight (1992), 96 of the 241 species (40%) are nearctic-neotropical or long-distance migratory species; 74 species (31%) are temperate migrants that have some populations wintering into the Neotropics; 4 species (2%) breed primarily along or south of the U.S./Mexican border; and the remaining 67 species (27%) are residents or migrants not defined by the Partners in Flight list.

In addition to being rich in migratory species, the riparian habitats of the middle Rio Grande support high densities of migratory birds. Densities of over 1,000 birds/100 acres (2,200 birds/km^2) have been recorded in cottonwood-willow habitats of the Rio Grande (Freehling 1982, Hink and Ohmart 1984, Hoffman 1990). High bird abundance and species richness in the middle Rio Grande valley are consistent with data reported for other riparian ecosystems in the Southwest (Hubbard 1971, Carothers et al. 1974, Ohmart and Anderson 1982; Rosenberg et al. 1982, 1991). Similarities in patterns of migration use along different southwestern drainages indicate the high value of these unique and limited habitats to landbirds migrating through semi-arid environments.

SEASON AND HABITAT EFFECTS

In 1981 and 1982, Hink and Ohmart (1984) conducted a biological survey along the middle Rio Grande. To evaluate whether bird abundance and species richness during spring and fall migration differed from abundance and richness in winter and summer, we reorganized and reanalyzed unpublished annual bird count data collected from 78 transects representing 21 different vegetation and structure types (see Hink and Ohmart 1984:104). We tested the effects of habitat type (cottonwood-willow, cottonwood-Russian olive, Russian olive, drainage habitat, and cattail marsh), vegetation structure (6 classes ranging from high to low foliage volumes in combinations of lower, middle, and upper strata; Hink and Ohmart 1984), and season (spring, summer, fall, and winter) on species richness and average density of birds. The results of our three-way analyses of variance showed that habitat type, vegetation structure, and season influenced species richness ($F_{12,63}$ = 5.43, P ≤ 0.001) and avian density ($F_{12,63}$ = 2.62, P = 0.008) in different ways. Species richness varied among seasons ($F_{12,3}$ = 10.12, P ≤ 0.001) with more species observed during spring and fall when migratory species move through the middle Rio Grande than in summer and winter (Fig. 3). Species richness also varied among habitat types ($F_{12,4}$ = 6.97, P ≤ 0.001) with more species detected in cottonwood-Russian olive and drainage habitats than in pure Russian olive or cottonwood-willow (Fig. 3). However, richness did not vary with vegetation structure ($F_{12,5}$ = 1.56, P = 0.187).

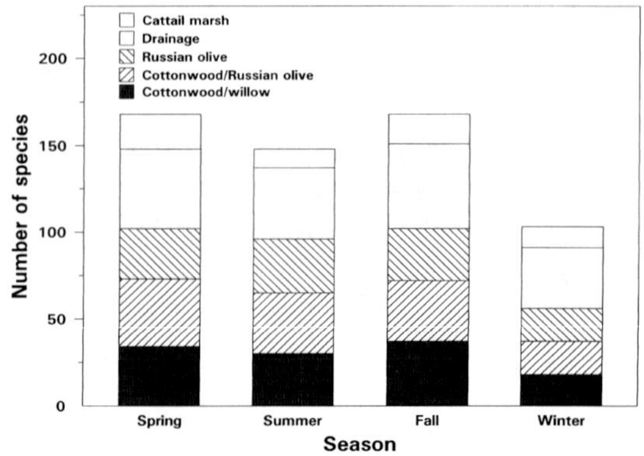

FIGURE 3. Species richness in relation to habitat type and season.

In contrast to species richness, bird densities reported by Hink and Ohmart (1984) were not influenced by season ($F_{12,3}$ = 1.50, P = 0.226), but were affected by habitat type ($F_{12,4}$ = 0.02) and vegetation structure ($F_{12,5}$ = 32.97, P = 0.02). Birds were more abundant at sites with intermediate-aged cottonwood trees and thick understories of willow or Russian olive; these formations occurred primarily along levees and river edges.

POPULATION TRENDS OF MIGRATING BIRDS

Although native riparian habitats have clearly changed since European settlement along the Rio Grande, long-term population data for different avian species are needed to determine whether habitat changes have affected migrating landbirds. Previous studies, including the long-term Breeding Birds Survey (BBS) coordinated by U.S. Geological Survey, Biological Resources Division, focused on breeding birds and offer little information on migration populations, habitat use, and effects of habitat alteration on landbirds that migrate through riparian systems in the western United States. Thompson et al. (1994) suggested that most bird species that were historically documented in the Rio Grande valley are still present, with the exception of Purple Martin (*Progne subis*), Red-headed Woodpecker (*Melanerpes erythrocephalus*), and Hooded Oriole (*Icterus cucullatus*). However, their historical information was qualitative, and population data were not available to be compared. A species could experience substantial population declines or increases over time while continuing to be recorded as present. We must be cautious in using records of continued species presence over time as indicative of the stability or well-being of a population. Over the short-term, evaluations of demographic data focused on nesting success, fledgling survival, and recruitment rates among local populations, habitats, and years may offer insight into the population status of a species.

We hypothesize that the habitat changes recorded along the Rio Grande (Fig. 2) have affected bird populations both positively and negatively, whether the populations are transitory, resident, wintering, or breeding. Some populations may have increased in response to changes in local conditions that have promoted increased availability of preferred habitats or structural types. Others have probably declined owing to invasion of alien plant species and associated increases in midstory shrub structure, expansion of urban areas, aging of cottonwoods, loss of marshes, and range expansion of cowbirds (Howe and Knopf 1991, Mount et al. 1996, Schweitzer et al. 1996). Because the middle Rio Grande is used by migrating birds that breed over a broader geographic area of western North America than that covered by the river itself (Yong and Finch 1996), alteration or loss of riverine stopover habitat could potentially have a large-scale effect on landbird populations.

Our hypothesis was supported by analyses of banding data collected by Rio Grande Bird Research, Inc. (RGBR; Yong and Finch 1997b).

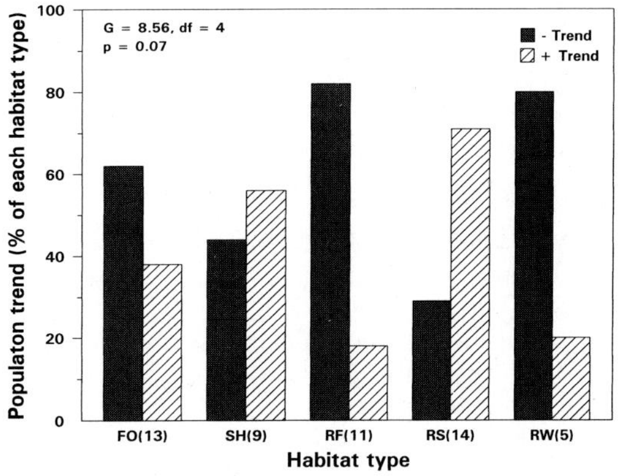

FIGURE 4. Relationship between population trends and breeding habitat types: FO = nonriparian forest, SH = nonriparian shrub/scrub/grassland, RF = riparian forest, RS = riparian shrub, and RW = riparian wetland, backwater, or marsh. Sample size of species is in parentheses along *x*-axis.

RGBR is a nonprofit group of volunteers who have banded fall migrants on weekends at Rio Grande Nature Center (RGNC; Fig. 1), within the city boundary of Albuquerque, New Mexico (35°07′N, 106°41′W), since 1979. The banding site is a riparian habitat that included bosque, two man-made ponds, and agricultural fields. We analyzed population trends of migrating landbirds based on data collected between 1985 and 1995. We excluded data collected prior to 1985 because the sampling methodology varied from year to year. We assigned migrating species to habitat use categories based on which habitats they most typically breed in and calculated what proportion of species showed negative or positive trends within each habitat category: upland forest, upland nonforest, riparian woods, riparian shrub, and wetlands. Trend direction was assigned based on slope calculated from regressions of birds captured/100 net-hrs with year as the predictor variable.

The most obvious pattern of population change was in riparian bird species, with 81% of shrub-using riparian species tending to increase and 29% decreasing; 82% of forest-dwelling species tending to decrease in population versus 18% increasing; and 80% of wetland species showing decreasing populations versus 20% having increases (Fig. 4). Examples of riparian forest birds that had negative population trends included Western Wood Pewee (*Contopus sordidulus*), Black-Headed Grosbeak (*Pheucticus melanocephalus*), and Warbling Vireo (*Vireo gilvus*). Riparian shrub dwellers with positive population trends included Blue Grosbeak (*Guiraca caerulea*), Lazuli Bunting (*Passerina amoena*), and Lincoln's Sparrow (*Melospiza lincolnii*). Common Yellowthroat, Black Phoebe (*Sayornis nigricans*), and Red-winged Blackbird (*Agelaius phoeniceus*) were some of the wetland species with negative trends.

STOPOVER ECOLOGY

Banding and Counting Study

Passerine birds may preferentially select riparian habitats as stopover sites during migration in the Southwest (according to Stevens et al. 1977, Hehnke and Stone 1979). During spring and fall migration, riparian systems can attract more than 10 times the number of migratory birds than surrounding upland sites (Steven et al. 1977, Hehnke and Stone 1979). Little information is available, however, on migrant stopover behavior and biology within riparian habitats during migration. To investigate timing and pattern of migration, migrant use of native, introduced, and agricultural habitats, and energetic condition and foraging behavior of migrants, we began a study of stopover birds in the middle Rio Grande. In spring 1994, we established a new banding station at Bosque del Apache National Wildlife Refuge (BNWR, N33°48′ and W106°52′; Fig. 1), located about 90 miles south of Albuquerque, New Mexico, and we expanded the existing banding program at RGNC (see above; Yong et al. 1995). Twenty nylon mist-nets (12 m × 2.6 m with 30- or 36-mm mesh) were used at each site to capture migrants during spring and fall migration. We also established 16 transects in the five dominant vegetation types: cottonwood, saltcedar, screwbean mesquite, willow transects, and agriculture lands. Each transect was 1 km long with point count stations located at 200-m intervals (6 stations/transect). Nets and transects were sampled every weekday.

Seasonal and Habitat Use During Migration

A total of 6,471 individuals of 108 species was captured at the two sites (data combined) during the 1994 field seasons. We captured 18 more species in the fall (93 species total) than in the spring (75 species). In addition, far more individuals (5,615 total birds) were captured during fall migration than in spring (856 birds; Fig. 5). Although mist-netting effort was constant between fall and spring with respect to numbers of net sites and netting hours per day, the number of fall migration days exceeded the number of migration days in spring, which may partially explain why more birds were captured in fall. In addition, hatching-year birds contribute to higher total numbers of fall migrants than spring migrants. More birds were detected from

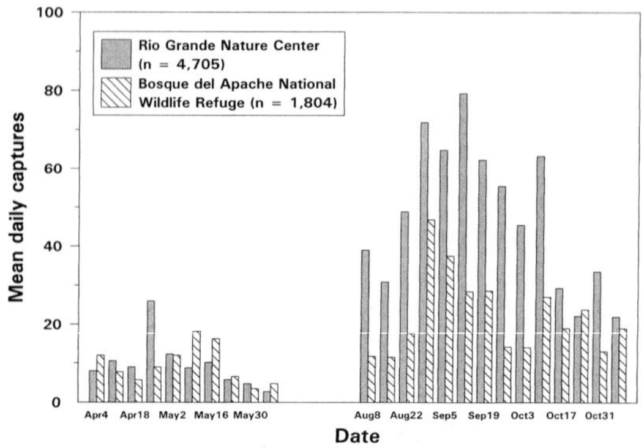

FIGURE 5. Seasonal capture pattern of landbird migrants at Rio Grande Nature Center and Bosque del Apache National Wildlife Refuge, New Mexico during spring and fall 1994. Daily averages are calculated based on weekly total captures.

mid-April to mid-May during spring migration and between early September to mid-October in fall migration than at other times. Our capture data for two seasons conflict with results of analyses from Hink and Ohmart's (1984) transect data, which suggested no seasonal variation in bird densities along the middle Rio Grande. Our data may not be directly comparable to Hink and Ohmart's data, however, because their intensive study area emphasized habitats north of ours and their methods of detecting birds differed from ours.

Numbers of landbird migrants captured and counted at BNWR and RGNC during spring migration varied among habitat types, suggesting that migrants differentially selected habitats during stopover. Given that we sampled a greater variety of habitats using transects than mist nets, we evaluated our transect data to detect any additional differences in stopover use among habitats. Based on results from our transect data for all migrants combined, 21% of migrants were observed in cottonwood-willow habitats, 25% in mesquite, 18% in saltcedar, and 35% in agricultural fields. Some species were observed in specific stopover habitats more frequently than others. For example, at the BNWR site surveyed in spring 1994, the relative distribution of Black-headed Grosbeak was 50.6% in cottonwood habitat, 16.1% in mesquite, 24.1% in saltcedar, and 8.0% in cropland, while the closely-related Blue Grosbeak was 15.79%, 19.55%, 19.55%, and 45.11%, respectively (Finch et al. 1995). Comparing stopover use of habitats by grosbeaks to habitat availability as approximated by 1989 National Wetlands Inventory data (Fig. 2), Black-headed Grosbeaks apparently selected forest types more frequently than they were available while avoiding agricultural fields despite their

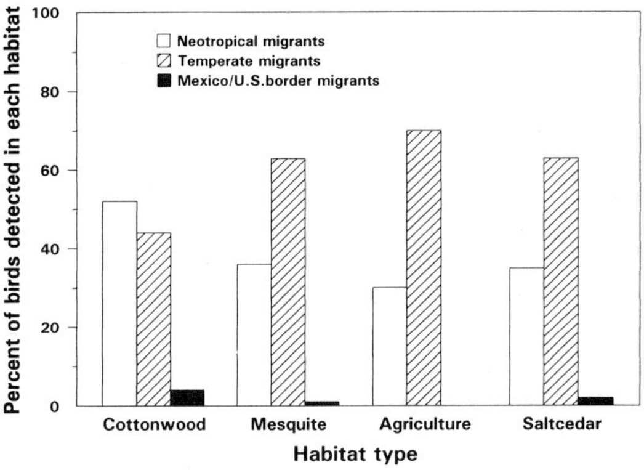

FIGURE 6. Proportions of total migrants observed in cottonwood-willow, mesquite, agriculture, and saltcedar in relation to Partners in Flight (PIF; Partners in Flight 1992) migratory status. Proportions are based on survey data from Bosque del Apache National Wildlife Refuge, spring 1994.

dominant presence. In contrast, cropland stopover use by Blue Grosbeaks more closely reflected cropland availability.

Nearctic-neotropical migrants used native cottonwood-willow habitat more often than other habitats in spring and more than temperate migrants (Chi-square = 18.36, df = 6, P = 0.005; Fig. 6). In contrast to findings based on Hink and Ohmart's data (Fig. 4), our current study suggests that more birds use native cottonwood-willow during stopover than habitats dominated by an exotic plant species, in this case, salt cedar (Fig. 6). For some bird species, this may reflect availability of habitats at local sites, but for certain tree-affiliated species (e.g., Black-headed Grosbeak, Western Wood Pewee) cottonwood forests are selected more often during stopover than would be expected based on availability (D. Finch and W. Yong, unpubl. data).

Differences in local availability, selection, age and structure of salt cedar (surveyed at BNWR) versus Russian olive (a dominant exotic in Hink and Ohmart's study) may influence stopover frequency in cottonwood-willow habitats. Birds may frequent cottonwood-willow habitats more than salt cedar when the former is more available at a site or if it provides more abundant or suitable resources, but may switch to Russian olive habitats in areas where olive is a prominent plant species. In the absence of cottonwoods, willows, and olives, monotypic stands of salt cedar may be the only wooded habitat available for riparian-dwelling bird species along some southwestern drainages (Livingston and Schemnitz 1996). Similarly to Hink and Ohmart's data for drainage habitats (Fig. 4), our unpublished mist-netting results suggest that even the recently established native coyote willow stands along drainage channels provided habitat for many nearctic-neotropical migrating landbirds such as Willow Flycatchers, Yellow-breasted Chats, MacGillivray's Warblers (*Oporornis tolmiei*), and Yellow Warblers.

Stopover Time Length and Fat Gain

We evaluated whether fat deposition and stopover length varied with time of day, species, or migration distance using methods defined by Moore and Kerlinger (1987). Such variation may reflect the relative importance or use of Rio Grande riparian habitats as refueling sites during migration. We selected Chipping Sparrow (*Spizella passerina*) to evaluate weight change over time because it was the most abundant species in fall 1994, thus providing a large sample size for detecting small changes in mass. Chipping Sparrows caught late in the banding morning weighed significantly more than sparrows caught early, with adults showing a higher rate of weight gain (0.12 g/hr) than hatching year birds (0.08 g/hr; Fig. 7). Weight gain over the morning suggests that birds were using stopover habitats to actively forage and replenish energy stores during migration.

We found mixed daily weight-gain results among species as well as intraspecific differences in weight gain by sex, age, season, habitat, and locality that we plan to publish in depth elsewhere. For example, female Wilson's Warblers (*Wilsonia pusilla*) showed significant (P > 0.05) mass gains of 0.16 g/6 hr to 0.36 g/6 hr in agricultural and cottonwood habitats at the RGNC during fall migration but not at the BNWR. In contrast, male Wilson's Warblers in fall showed significant weight gains of 0.20 g/6 hr to 0.32 g/6 hr in various cottonwood habitats at both RGNC and BNWR but not in agricultural or willow habitats. In some species, hatching year birds had lower and more variable weight gain than adults, possibly because they were less experienced and less skillful in finding and consuming food during fall migration through unfamiliar habitats (Woodrey *this volume*).

Our fall 1994 banding data from RGNC and BNWR suggest that stopover time is dependent on whether specific species need to replenish energy stores. Species that stopped in habitats during our morning netting periods most frequently exhibited zero or very slight fat stores (Fig. 8), suggesting that the stopover was needed for refueling. Numbers of nearctic-neotropical migrants did not differ from temperate migrants among five fat classes (Chi-Square = 2.88, df = 4, P = 0.58) (Fig. 8), although in comparing the first two fat classes only, long-distance migrants

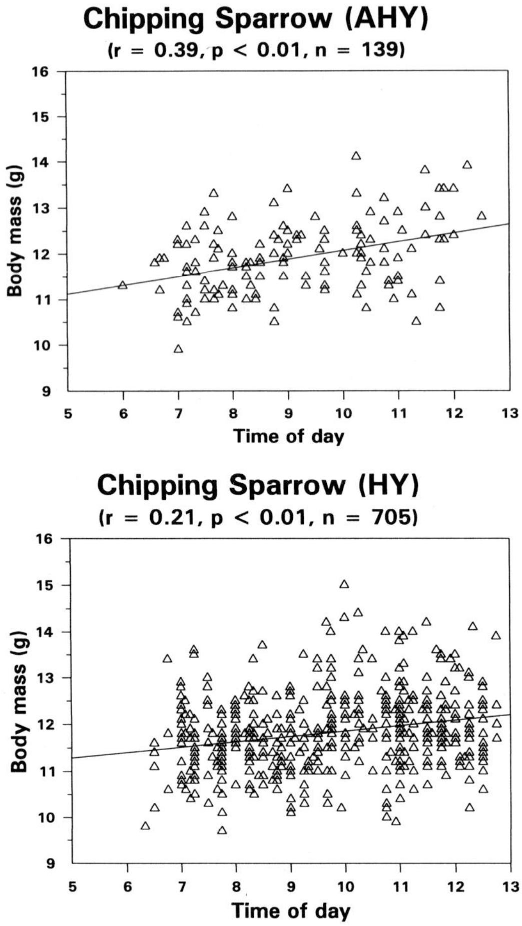

FIGURE 7. Weight change (g) over time in adult (AHY) and hatching year (HY) Chipping Sparrows captured during morning mist-netting sessions in fall, 1994.

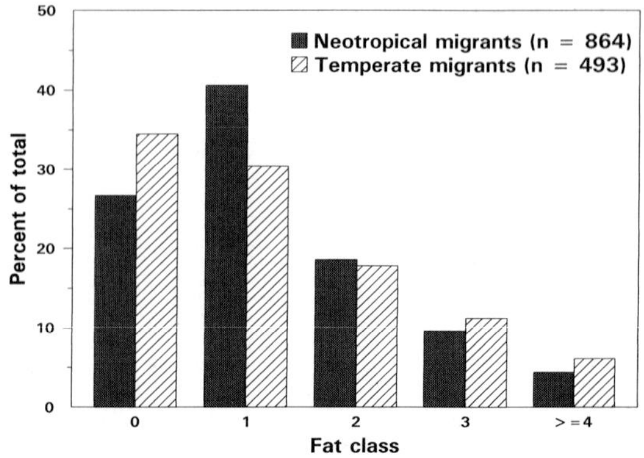

FIGURE 8. Fat class distributions of nearctic-neotropical (PIF A) and temperate (PIF B) migrants captured at the Bosque del Apache National Wildlife Refuge and Rio Grande Nature Center in spring 1994. Fat classes range from 0 to 5, with 0 = no fat to 5 = very fat (sensu Moore and Kerlinger 1987).

appeared proportionately more abundant in Fat Class One than temperate migrants (Chi-square = 2.42, df = 1, P = 0.12; Fig. 8). Perhaps long-distance migrants prepare for longer flights by storing more fat than temperate or short-distance migrants. For those species crossing the Chihuahuan Desert during spring migration, the bosque of the middle Rio Grande may serve as an especially important stopover site for refueling.

Analyses of recapture data for three transient species, Dusky Flycatchers (*Empidonax oberholseri*; nearctic-neotropical) MacGillivray's Warblers (nearctic-neotropical), and Hermit Thrushes (*Catharus guttatus*; temperate), showed that they had relatively short stopover lengths (1.5 days, 1.75 days, and 2.71 days, respectively) and relatively large amounts of individual mass (fat) gain (4.07%, 13.44%, and 3.01%, respectively) on average (Fig. 9). In contrast, Blue Grosbeak, a nearctic-neotropical migrant that breeds locally in the middle Rio Grande bosque, had a longer average stopover length (11 days) and smaller average mass change (−0.15%; Fig. 9). These preliminary data, while limited in sample size, support the idea that transients fortify themselves during stopover more than do migrants that are close to or within the vicinity of their breeding grounds.

DISCUSSION AND RECOMMENDATIONS

Based on analyses of 1984 bird count data and more recent bird capture data, we conclude that species richness and abundance of birds are greatly influenced by season and habitat type along the Rio Grande. Our mist-netting results showed that season influenced overall bird abundance more than the results of our analyses of Hink's and Ohmart's 1984 transect data suggested. Our comparisons of spring and fall bird-banding data demonstrated strong differences in abundances between these two seasons alone. The transect method used by Hink and Ohmart may not adequately sample some bird species in migration because migrating birds are less vocal than breeding birds, more transient than both breeding and wintering birds, and more difficult to identify in spring, fall, or hatching-year plumages. In addition, many transient species are rare or accidental. Therefore, migrating birds are often difficult to detect using traditional counting methods. We recommend that birds in migration be sampled using mist-nets in association with other counting methods to estimate abundance and species presence. Nevertheless, traditional counting methods such as point counts and tran-

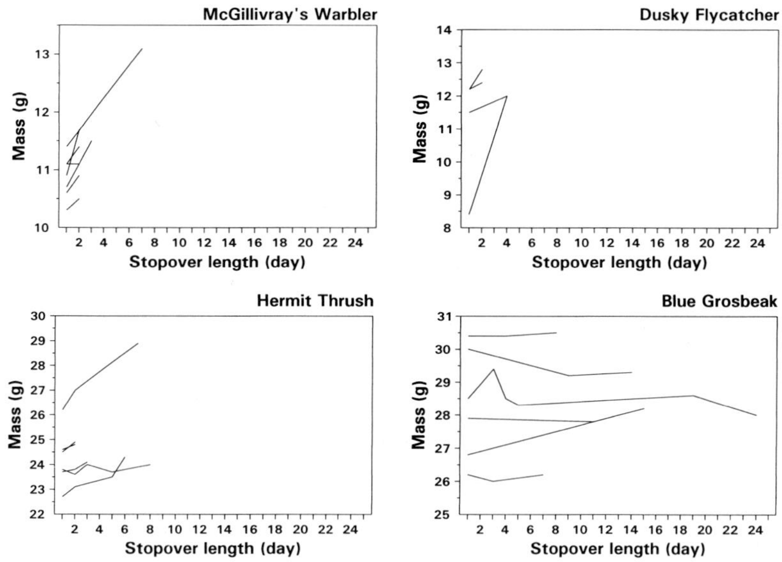

FIGURE 9. Body mass (g) gain in relation to length of stopover time based on recapture data for MacGillivray's Warbler, Dusky Flycatcher, Hermit Thrush, and Blue Grosbeak in fall, 1994. Each line represents an individual bird.

sects are recommended as the most efficient, economical means for sampling relative abundance of populations among habitats and species over a broad number of sites. A species-by-species or guild-level analysis of Hink and Ohmart's data may further clarify patterns of abundance and species richness among habitats and seasons.

Our analysis of Hink and Ohmart's data documented that Russian olive, an alien plant species, was used to a considerable extent by birds year round. Its olive crop provides a food source to many bird species, and its structural form offers cover and nest substrate to understory birds (Van Dersal 1939, Freehling 1982). Questions have been raised as to whether this introduced woody species has added value to the system (Freehling 1982), or whether it should be viewed as a disturbance feature that should be eliminated or controlled. This question is difficult to answer because of lack of information on bird populations and habitat use of these systems under pre-Russian olive conditions. Teasing out whether specific bird species are closely tied to or avoid Russian olive habitats in relation to arthropod supplies or nutritional value of olives, and in the presence and absence of cottonwood-dominated overstories, may help to quantify precisely whether and how Russian olive adds value, and whether this value offsets the disturbance factor.

The same reservation holds true for salt cedar habitats, which are reported to differ in bird species composition from native habitats (Hunter et al. 1988, Farley et al. 1994, Ellis 1995). However, when mixed with other woody plants, salt cedar habitat is reported to be more valuable to landbird species along the Rio Grande than monotypic or manipulated vegetation (Leal et al. 1996). Along certain New Mexico drainages such as the Pecos River, salt cedar offers new wooded habitat where few woods were historically available (Livingston and Schemnitz 1996). Even so, habitats dominated by such alien woody plants may represent an ecological trap or sink if populations of some of the bird species that use them are unable to sustain productivity. In addition, if other habitats such as grasslands or marshes are gradually being displaced by invasive alien plants, some of the bird species associated with the displaced habitats are likely being lost from sites also (Livingston and Schemnitz 1996). We recommend studies that compare breeding success of species' populations among habitats dominated by alien and native plant species. Habitat use of introduced plants by nearctic-neotropical migratory bird species that have been identified as high priorities for conservation (e.g., Hunter et al. 1993) should be assessed and closely monitored. If priority bird species are positively or negatively associated with exotic plant species, then conservation action may be important (Leal et al. 1996).

While our analyses of local population data collected between 1985 and 1995 from the Rio

Grande Nature Center in Albuquerque were limited in sample size and did not explain whether populations changes were associated with events on the breeding grounds, wintering grounds, migratory routes, or a combination of these, they did suggest that selected species that use the Rio Grande during fall migration have experienced changes in numbers over a period of ten years. In addition, numerical increases and declines of riparian-breeding species tended to separate out into broad habitat categories. Changes specific to riparian shrub, riparian forest, and wetland species may be related to western expansion of exotic woody species resulting in increased cover of shrub understories and mixed overstories along rivers and streams (e.g., Mount et al. 1996, Leal et al. 1996), aging and dying of cottonwoods in many western riparian systems (Howe and Knopf 1991, Finch et al. 1995), and widespread loss of western marshes and wetlands (Mitsch and Gosselink 1986, DeSante and George 1994, Noss et al. 1995). Population changes that we observed may or may not be directly linked to local habitat changes along the Rio Grande, but certainly the Rio Grande is symptomatic of, and contributes to, problems in western riparian and wetland systems.

Although our stopover data were preliminary, they suggest that riparian woodlands along the middle Rio Grande are valuable to numerous species of migrating landbirds. As our fat deposition results demonstrated, migrating birds use the bosque to replenish energy stores during travel. We postulate that stopover along the middle Rio Grande may be especially important for those species whose migration flights cross the Chihuahuan Desert of Chihuahua, Mexico, southern New Mexico, and western Texas. With regard to the need for depositing fat reserves, migration over vast deserts may be analogous to flights over large water bodies such as the Gulf of Mexico (e.g., Moore and Kerlinger 1987, Moore et al. 1995). Because most riparian woodlands along the Rio Grande in New Mexico are north of Elephant Butte Reservoir, some spring migrants may travel for a considerable distance across inhospitable desert, possibly following the river channel, before reaching the bosque of the middle Rio Grande. Survival of some individuals could conceivably depend on reaching this woodland resource before energy stores are completely depleted. In fall, survival of inexperienced hatching year birds that stop in the bosque on their first flight south could hinge on how much fat they have deposited before departure over desert country. When the Rio Grande is viewed in such a light, the need to conserve its bosque becomes as obvious as the bosque's importance to birds.

Variation in the abundance distributions of migrating species suggest that Rio Grande habitat types differ in value among species (for breeding birds, see also Leal et al. 1996). Thus, changes in habitat cover, structure, and composition may influence habitat use, food availability, health, and survival during migration, thus potentially influencing population trends (e.g., Moore and Simons 1992a). Factors that reduce the suitability of riparian habitats as stopover sites in the middle Rio Grande valley may affect not only local birds, but also populations at larger geographic scales, and may conceivably jeopardize the future of some populations. Riparian woodlands along other major drainages of the Southwest appear similarly important to transients and locally-breeding migrants (e.g., the lower Colorado River; Rosenberg et al. 1991). Conservation of the Rio Grande bosque has already been elevated to a high priority based on its high biological diversity (Crawford et al. 1993). We concur and further recommend that migration habitats be recognized and included as an important factor in the planning and design of conservation and restoration projects not only for reaches along the middle Rio Grande, but for riparian vegetation along southwestern rivers and streams in general.

ACKNOWLEDGMENTS

We thank Rio Grande Bird Research, Inc., for providing banding data from the Rio Grande Nature Center for our analyses. We are grateful to the staff at the Rio Grande Nature Center and Bosque del Apache National Wildlife Refuge for access, housing, and assistance with our migratory bird studies at these two sites. We thank V. C. Hink, R. D. Ohmart, and Army Corps of Engineers for permission to use their unpublished report. Many technicians and volunteers collected data in the field, and we appreciate their help. In particular, we acknowledge G. Bodner, D. Hawksworth, J. Sevigny, and M. Sevigny for their stalwart field assistance. We are especially indebted to Frank Moore for supporting a cooperative agreement between the Rocky Mountain Experiment Station and University of Southern Mississippi to finance W. Yong's visiting postdoctoral position in Albuquerque. We thank two anonymous reviewers for their thoughtful comments on the manuscript.

Studies in Avian Biology No. 20:99–108, 2000.

CONSERVATION OF LANDBIRD MIGRANTS: ADDRESSING LOCAL POLICY

SARAH E. MABEY AND BRYAN D. WATTS

Abstract. Proactive conservation measures on behalf of neotropical migrants are gaining strength and legitimacy within government agencies and private conservation organizations throughout the Western Hemisphere. Most of these efforts focus on managing large tracts of public and private land or acquiring land for outright preservation. These strategies do little to confront threats facing the vast aggregate of relatively small, private land parcels. Taking conservation beyond the boundaries of public land requires the use of policy and management tools not conventionally tied to ecological issues. Northampton County, Virginia, located on the lower Delmarva Peninsula and home to large numbers of migrant landbirds every fall, provides an example of a local community testing the application of such tools to the global problem of migratory bird habitat protection. The county's Special Area Management Plan (SAMP), funded under the Coastal Zone Management Act, identifies protection of migratory bird habitat as an essential element in fostering a sustainable local economy and mitigating the secondary impacts of coastal development. The SAMP was designed to involve a diverse group of local, state, federal, and private partners and has included sponsoring a two-year research project on the local geographic and ecological distribution of fall migrants and recruiting public support through ecotourism initiatives. SAMP partners are currently applying the results of the migrant/habitat research to zoning ordinances and various memoranda of understanding that address local habitat protection. Based on the example of Northampton County, we suggest that local communities may be willing to apply land-use policy to stopover habitat protection if scientists provide them with information necessary for conservation planning.

Key Words: conservation, land use, landbird migrants, local policy.

The papers in this and other volumes (Hagan and Johnston 1992, Finch and Stangel 1993, Martin and Finch 1995) highlight the special challenges faced in understanding and conserving neotropical landbird migrants. Over the past decade, numerous state, national, and international migratory bird conservation programs have been established, many of which are coordinated through the National Fish and Wildlife Foundation's Partners-in-Flight initiative. These programs reflect current scientific knowledge and represent frontiers in conservation. Until recently, however, many of these programs have failed to fully incorporate issues related to stopover ecology and exploit possibilities for conservation initiatives at a local level (Greenberg and Lumpkin 1991, Johnson 1993, Mabey et al. 1993, Wigley and Sweeney 1993, Watts and Mabey 1994).

The challenge of conserving stopover habitat for landbird migrants differs from that of protecting breeding and wintering habitats in at least two critical ways: habitat heterogeneity and scale. Habitat heterogeneity poses an ecological and energetic dilemma for birds, whereas scale is a political and economic challenge for human conservation efforts. During the course of migration, an individual migrant moves through an extremely heterogeneous environment. The relative quality of habitats within that matrix will directly influence the bird's ability to complete migration (Simons et al. *this volume,* Parrish *this volume*) and may indirectly affect its survival or breeding success. Although migration ecology is an expanding field (Crick and Jones 1992, Hagan and Johnston 1992, Moore et al. 1995; chapters in *this volume*), relationships between migrating birds and their environment remain inadequately understood.

The spatial scale of migration presents difficulties with respect to assigning responsibility for the protection of a population or species of landbird migrants; it would appear that federal responsibility is necessary. A well-coordinated policy might cover public land across the country, creating a continental safety-net. In fact, Partners-in-Flight has developed sound, science-based management objectives for public lands and created the coalition of governmental agencies necessary to attain those goals (Finch and Stangel 1993).

However, as Wigley and Sweeney (1993) have argued, a safety-net of public lands is insufficient to confront the two problems of habitat heterogeneity and scale. Within the United States, the Federal Government manages 649.8 million acres of land, nearly 29% of the country's land mass, over 56% (367.6 million acres) of which is maintained for forest and wildlife usage (US General Services Administration 1993). Federal lands are not, however, evenly distributed. They are highly concentrated in the western states and account for less than 5% of the area of the eastern states (Fig. 1). This is a

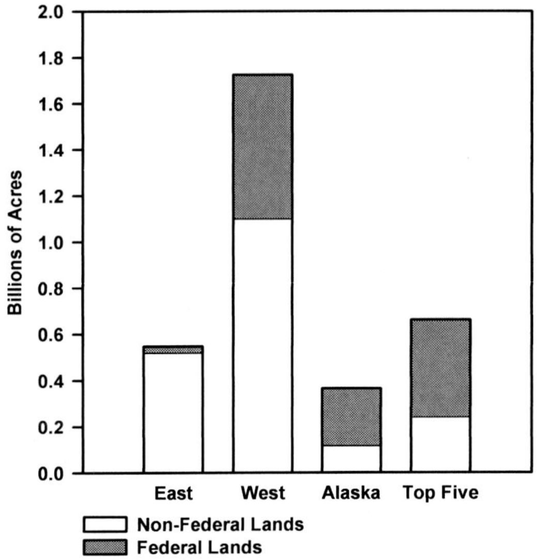

FIGURE 1. Distribution of federally managed lands within the United States. East includes all states east of the Mississippi River and West includes those to the west. Top Five includes the five states containing the largest acreage of federal lands (Alaska, Nevada, California, Arizona, and Utah) (US General Services Administration 1993).

potential problem as the majority of neotropical migrant species and individuals migrate east of the Rocky Mountains (Moore et al. 1995). Although a small percentage of the remaining land is held by state governments or private conservation organizations for the purpose of land protection, more than 90% of the land in the eastern United States remains in the hands of private landowners.

Unfortunately, it is within this expansive realm of private property that habitat degradation is most severe. Areas offering minimal stopover support to neotropical migrants are rapidly spreading across the continent with acute deterioration along the coasts. Few federal (e.g., the Endangered Species Act) or state laws regulate activities that affect natural resources on private property. The well-publicized controversy surrounding the Endangered Species Act (ESA) highlights some of the limitations to federal authority on private property (Dwyer et al. 1995, Bean and Wilcove 1997), particularly the tension between local and national public interests (Mangel et al. 1996, Press et al. 1996). However, since the law was amended in the early 1980s, effective applications of the ESA have begun to emphasize local, long-range planning involving both private landowners and surrounding communities through the use of Habitat Conservation Plans (HCPs). In most cases, landowners now have the option of altering critical endangered species' habitat in a given area if it is possible to mitigate the effect of the development (Dwyer et al. 1995, Bingham and Noon 1997). Habitat conservation plans generally involve low-impact development designs that leave at least some critical habitat intact, or land trades requiring the purchase and protection of comparable habitat in another location. Although HCPs provide landowners greater flexibility in dealing with endangered species on private property, they are usually difficult to negotiate and offer only piecemeal protection of critical habitat (Bean and Wilcove 1997).

Despite the difficulties of protecting critical habitat for endangered and threatened species on private property through the federal regulation, the ESA and HCP process exists and can serve as a basis for discussion and compromise. There is no comparable regulatory protection process for migrant stopover habitat. Yet, decisions made by private landowners have a crucial impact on the future of neotropical landbird migrants and their habitats. In turn, such decisions are strongly influenced by the local economic, social, political, and regulatory climate. For this reason, local initiatives are imperative to the success of any comprehensive conservation plan for landbird migrants.

With this paper we call attention to the utility and strengths of applying local land-use policy and other locally-driven initiatives to the challenge of migrant stopover habitat protection. Although the unique aspect of local conditions limits the general relevance of a case study approach to understanding conservation through local land-use policy, there is value in examining the successes and difficulties of applying local land-use regulation to the protection of migrant stopover habitat in a real community. Local land-use regulations reflect immediate community standards and priorities. They represent small populations and, if approved, often have a better chance of success than federal or state level regulations. We present an overview of regulatory and voluntary methods frequently employed for the protection of natural resources on private property. A working example from Northampton County, Virginia, serves as an illustration of the value of community-based, community-focused initiatives for the conservation of neotropical landbird migrants.

PROTECTING HABITAT ON PRIVATE LAND

The problems involved in protecting resources for a dynamic, mobile, and somewhat unpredictable group like migrating landbirds run parallel to those encountered by the current movement to protect entire ecosystems (Carroll and Hendrix 1992). Closing individual parcels of land to the public will not address all of the real and potential threats to the resource. Working within a broader context that includes human communities and individual private landowners has taken on a new importance (Soulé 1991, Endicott 1993, Press et al. 1997). There are a variety of approaches to protecting natural resources on private property. Most fall within one of two basic categories of action: voluntary or regulatory. Voluntary land protection tools can be divided into six main areas: (1) acquisition; (2) easements; (3) natural area dedication; (4) management agreements; (5) government or private economic incentives; and (6) independent actions related to use or design (i.e., creating a natural landscaping plan, initiating ecotourism ventures, or opting for no human use).

Although acquisition offers the highest level of protection and has been by far the most frequently used tool, there are two serious constraints to its practicality. First, money for purchase must be raised either through private donations or dedication of tax dollars. The dimensions of this obstacle are determined by land prices and the level of interest among the citizenry. Additionally, local resistance to land purchase for conservation can be strong because there is often confusion regarding how such action might affect the local tax base. In some situations properties strictly dedicated to resource preservation can be removed from the local tax base, even though this negative is usually counter-balanced by increases in surrounding property values. An example from Northampton County, Virginia, illustrates the financial limits of acquisition. On Virginia's Eastern Shore, the most recent public land acquisition is Kiptopeke State Park. The park encompasses three hundred and ninety-five acres that had been readied for private development and cost almost \$28,000/acre. While half of the area is designated a natural area, the other half is devoted to crop production and recreational use. The commonwealth paid \$11 million to protect a little more than 1% of Northampton County. For comparison, a 1992 bond initiative passed by Virginia voters allocated only \$11 million for natural area acquisition for the entire commonwealth.

The second main problem with conservation land acquisition is related to the issue of ownership. Even if the interest in the conservation goal is strong and money can be raised to purchase land, someone must also take responsibility for the maintenance and management costs of the property, which may include liability insurance, security patrols, access improvements, and property taxes. Occasionally, political battles erupt over land ownership because local communities resent the intrusion of "outsiders" or because of the difficulties in forging partnerships involving local, state, federal, and private entities.

In light of these budgetary and political restrictions on acquisition, conservation efforts increasingly focus on other voluntary land protection tools (Endicott 1993). Easements and natural area dedication are legally binding contracts that can offer protection in perpetuity. Easements involve the sale or donation of some or all of the development rights associated with a piece of land. The landowner and easement holder agree to general management guidelines and restrictions that are incorporated into the title of the land. The landowner is compensated either directly (purchased easement) or indirectly (tax benefits from donated easement) for accepting development constraints on the property. The easement holder, either governmental or non-governmental, accepts the responsibility for enforcing and defending the easement, especially when the property is transferred to a new owner. Natural area dedication is a variant of the easement process usually involving a governmental organization and donation, rather than sale, of all development rights. Management agreements, in contrast, are good faith agreements between the current landowner and a conservation organization. They do not remain with the deed of the property and compensation is limited to management advice and the personal rewards of doing a good deed. The implementation of such voluntary measures require that the individual landowner has a relatively strong understanding of the ecological value of the property and a willingness to sacrifice in some way for the preservation of that value. This is particularly true in the case of voluntary, ecologically-sensitive development design when the landowner is making decisions based solely on an assessment of personal benefit.

Despite the strength and frequent use of voluntary land protection tools, regulatory actions are often a necessary complement within a local conservation strategy. Local level natural resource regulation may be incorporated into an array of land-use ordinances (e.g., zoning, cluster development, transferable development rights) or tax incentive programs. A local community may recognize a conservation issue through confrontation with outside interests or it might surface as an area of concern during a community's planning process. Resource protection problems brought to the fore by outside interests are not necessarily doomed to failure but may take longer to resolve. Likewise, those issues identified from within are not necessarily destined for successful resolution. However, conservation themes and specific problems incorporated into a community's comprehensive plan are certain to receive repeated attention until the plan's goals for protection are met. A comprehensive plan must be approved by voters and should represent the economic, social and aesthetic priorities of the local community. The comprehensive plan provides the fundamental justification for the creation of enforceable policy designed to protect resources. Development and implementation of resource conservation policies follow from the comprehensive plan. As with voluntary actions, the success of regulation relies heavily on two factors: the community's appreciation of and willingness to protect intact natural resources and its understanding of the economic implications of conservation.

In the United States, zoning is the most common form of directing different types of development to the most appropriate geographic areas within a community and controlling building density. Zoning is also used to formally express a community's common conservation and aesthetic values. Beyond zoning, local land-use ordinances can be designed to increase open space or protect special features of the landscape. Subdivision ordinances can promote cluster development by setting a low a maximum house lot size and maximum distance to nearest neighbors rather than a high minimum size and minimum distance while maintaining absolute housing density (number of houses per acre) allowed under the zoning law. This acts to group houses close together, leaving large areas of common open space.

Transferable development rights (TDRs) have a similar effect on a larger scale. In communities allowing TDRs, landowners may sell development rights for the maximum number and kind of building units permitted on their property. In doing so, the landowner erases these rights from the property title and they are added to the title of the buyer's property. Through TDRs, buyers can increase the maximum allowable building density on their properties but the overall community-wide density remains fixed. As with cluster-style development, use of TDRs may result in decreased demands on public services such as sewers and roads.

Unfortunately, the use of tax incentives for conservation of natural resources on private property is relatively rare at the local level. While some communities have special agricultural or silvicultural districts for the purpose of taxing land at current-use value, most local governments are uncertain of the economic and political repercussions of providing meaningful land tax breaks. As a result, the standard for property tax values is generally set by locally-determined "highest and best use" of the land. In rural and suburban communities, the "highest and best use" often translates into the value of the property at maximum building density. This standard has a profound effect on natural resources because it means that undeveloped land is taxed out of the bounds of economic viability for the landowner, who may be forced to develop or sell the property.

Most communities choose their set of conservation tools largely on the basis of the source and strength of their motivation to preserve and protect natural resources. A community expecting direct economic benefit from resource protection (e.g., from ecotourism) may find implementation of incentive programs financially feasible and voluntary conservation actions relatively easy to initiate. Alternatively, an awareness of severe costs from inaction (e.g., consequences of noncompliance with federal or state regulations) may lead a local community to take proactive regulatory measures, especially if there is a choice between locally- or federally-controlled standards. This principle is well-illustrated by the case of the California Gnatcatcher (*Polioptila californica*) habitat conservation planning process for San Diego and Orange counties, California. The California Gnatcatcher occupies coastal scrub habitat on some of the most expensive real estate in the United States and is listed as threatened under the Endangered Species Act. In this situation, either strict compliance or non-compliance with federal law would have translated into extreme costs for the local economy. Developers and local officials have had a strong inducement to face federal habitat protection regulations with a proactive compromise grounded in local priorities (Mann and Plummer 1995). The resulting Natural Community Conservation Plan combines voluntary and regulatory tools to protect coastal scrub habitat for the gnatcatcher and other rare plants and animals.

Occasionally, the local zoning ordinance process can work against habitat conservation. An example of a special feature ordinance inadvertently affecting stopover habitat recently occurred in Cameron Parish, Louisiana, where the Baton Rouge Audubon Society (BARS) manages the Henshaw Sanctuary. Henshaw protects coastal chenier habitat, and part of the BARS management plan for the property has been to allow for natural regeneration of native vegetation. Apparently in reaction to an unrelated conflict between sanctuary personnel and neighboring landowners, Cameron Parish decided to enforce a local weed control ordinance and, in the summer of 1996, ordered BARS to mow the

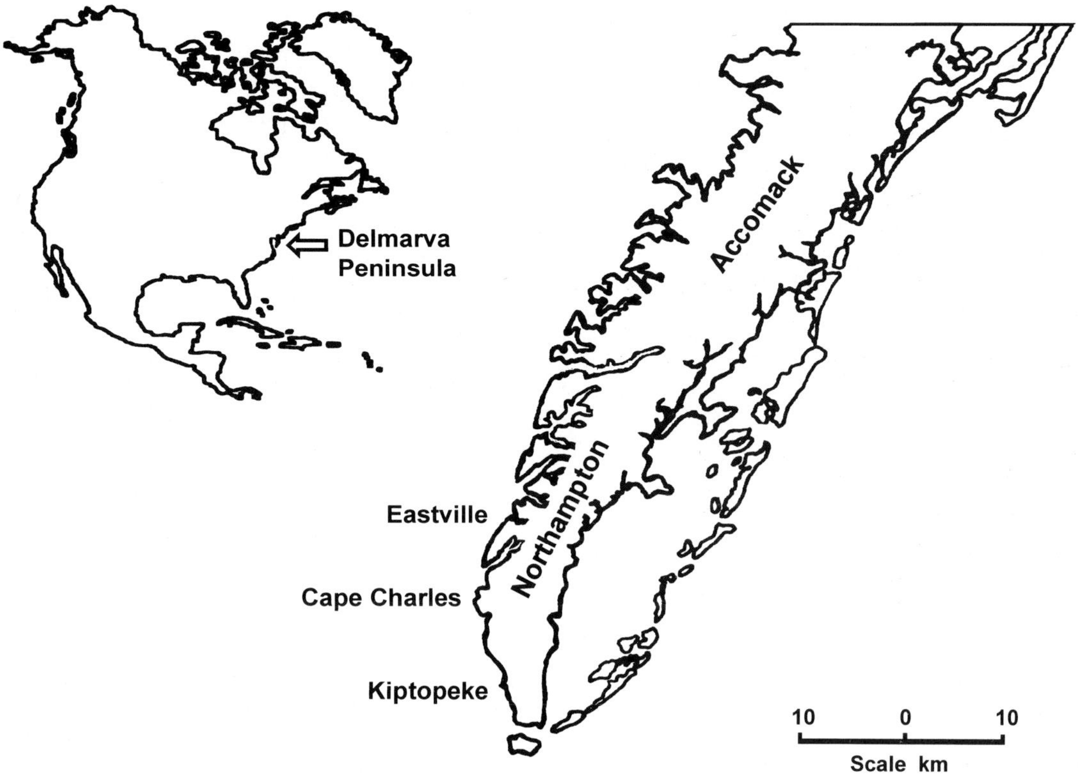

FIGURE 2. Location of Northampton County, Virginia.

sanctuary. BARS contested the order in court based on the vague language and arbitrary enforcement of the ordinance. BARS won their lawsuit in court and the sanctuary has been allowed a permanent exception to the ordinance (The Barred Owl 1996a,b). This case illustrates two important points: first, local land-use policy affects even those private landowners intent on habitat conservation; and second, land-use policy that is uninformed by science can be particularly dangerous to conservation efforts. The Cameron Parish ordinance made no distinction between good migrant stopover habitat and weeds. Such details are critical yet easily overlooked.

A WORKING EXAMPLE: NORTHAMPTON COUNTY, VIRGINIA

An on-going project from Northampton County, Virginia serves to illustrate the application of both regulatory and voluntary protection for migratory bird stopover habitat. Northampton County covers the southern 50 km of the Delmarva Peninsula, including the coastal barrier islands from the southernmost Fisherman's Island north to Hog Island (Fig. 2). Along the western coast, the landscape is dominated by the tidal creeks, maritime forests, and dune grass and scrub communities of the Chesapeake Bay; to the east lies a vast coastal wilderness of marshes, lagoons, and undeveloped barrier islands. The mainland is covered by moderate-sized farm fields, many still separated by hedgerows, and fragmented forest. Forests are primarily mixed pine/deciduous, but bottomland deciduous forest and loblolly pine plantations are also common. Marsh/upland ecotones and wetland forests stretch along the seaside. The area has been designated a United Nations International Biosphere Reserve.

This thin strip of land and neighboring islands have long been celebrated for their great abundance of migrating shorebirds and wintering waterfowl. Through the late 1920s, sport and harvest hunting for local consumption and the millinery trade contributed substantially to the local economy. Migrating landbirds have received much less intense but more benevolent attention. Rusling (1936) was the first to scientifically document hawk migration on the lower Delmarva. Since the early 1970s a group of volunteers has maintained the Kiptopeke Hawkwatch, each fall counting many thousands of migrating raptors. The Virginia Society of Ornithology established

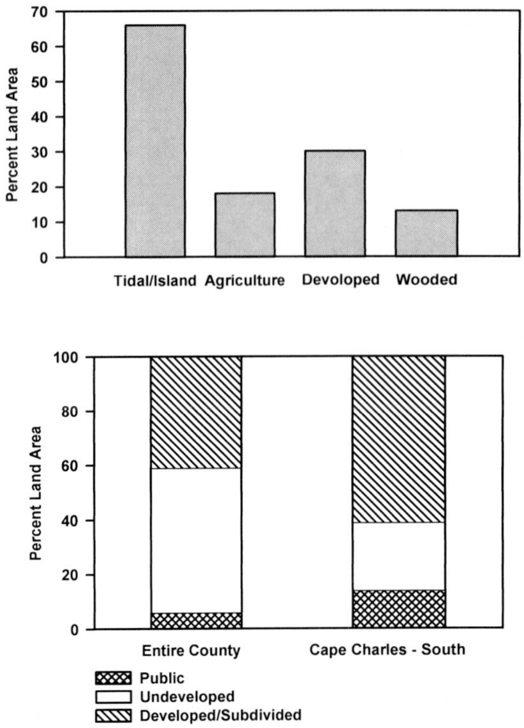

FIGURE 3. Land use patterns within Northampton County. A. Major land types and use in percent coverage, 1985. B. Ownership and use of bayside shoreline property for the entire county and the southern portion below Cape Charles, 1993 (Northampton County Department of Planning and Zoning, unpubl. data).

an "Operation Recovery" banding station at Kiptopeke in the early 1960s and continues to band thousands of south-bound migrants every year. Recent studies and reports further document the importance of this area for migrating landbirds (Armistead 1993, McCann et al. 1993, Mabey et al. 1993, Watts and Mabey 1994).

From the human perspective, Northampton County is a rural and economically depressed community of 13,000 residents. Northampton ranks 135th in poverty measures out of Virginia's 136 localities (Virginia 1990 Census Data). The largest town in the county, Cape Charles, has a population of under 1,500. Land-use patterns in the county have changed little in this century because much of the existing forest land is unsuitable for crop production. Farming is the dominant land-use, followed by silviculture, with relatively small areas developed for residential, commercial, or industrial use (Fig. 3a). Rapid change, however, is on the horizon. In the past eight years (particularly prior to the passage and implementation of the Chesapeake Bay Preservation Act), there has been a rush to subdivide waterfront property, especially on the bayside (Fig. 3b). A real estate recession has slowed real development, however, and as of 1995, nearly 4,500 platted lots stood empty (Northampton County Department of Planning and Zoning data).

Northampton County has received substantial conservation attention, and local, state, federal, and private entities have collectively employed most voluntary methods of land protection available. The barrier islands constitute the most sensitive portion of the ecosystem and are protected largely through ownership by The Nature Conservancy, the US Fish and Wildlife Service, the Virginia Department of Game and Inland Fisheries, and the Virginia Department of Conservation and Recreation. The Nature Conservancy's Virginia Coast Reserve owns and manages the majority of the barrier islands as the core area of their flagship bioreserve. On the mainland, land is protected by a US Fish and Wildlife Service National Wildlife Refuge, a state park and state natural area preserves, and wildlife management areas. Conservation easement activity is concentrated along the seaside of the mainland but can be found throughout the county. Regulatory measures are now being considered to fill in the gaps, especially along the bayside where the migrants concentrate, development pressure is high, and important areas are already heavily subdivided (Northampton County Department of Planning and Zoning 1989, Mabey et al. 1993, Watts and Mabey 1994).

Despite daunting socio-economic conditions and an already high degree of conservation activity, this community adopted a progressive comprehensive plan in 1990 that clearly states that the County must conserve its natural resources and specifically mentions migratory birds (Northampton County Joint Planning Commission 1990). Through its comprehensive plan and subsequent planning initiatives, Northampton County has demonstrated a commitment to taking a different direction from most of the rest of the Eastern Seaboard.

Northampton's primary motivation for accepting the challenge of stopover habitat protection is based on economic interests in the rapidly growing nature-based tourism industry (Citizens for a Better Eastern Shore Newsletter 1991). Birdwatching enthusiasts represent 14% of the American public and spend billions of dollars annually on birding excursions, equipment, memberships, and other related paraphernalia (Wiedner and Kerlinger 1990). Kerlinger and Wiedner's (1991) study of the economics of birdwatching indicates that birdwatchers spend over $5 million a year in Cape May, New Jersey.

Current estimates for Cape May's annual avitourism earnings are greater than $20 million (P. Kerlinger, pers. comm.). If Northampton County could build the eco-tourism industry to a similar level, it would place among the top five contributors to the local economy.

A regional study of migratory bird distribution along the coasts of the Cape May and Delmarva peninsulas demonstrated that the numbers of fall migrants in Northampton County were higher than those of Cape May (McCann et al. 1993). In light of the economic benefits birdwatchers bring to Cape May, this comparison drew the County's interest, and officials and citizens became receptive to the idea of sustainably capitalizing on migratory birds (Citizens for a Better Eastern Shore Newsletter 1991). In 1992 the opportunity to do so arrived in the form of a Virginia Coastal Resources Management Program grant funded by the National Oceanic and Atmospheric Administration (NOAA). This Special Area Management Plan (SAMP) grant has provided over $1 million during a four year period to develop "new and enforceable policies to *protect* and *enhance* coastal resources" (emphasis added). Congressional authorization for the SAMP funding program (Coastal Zone Act §309A) explicitly states that SAMP strategies should promote intergovernmental cooperation and control impacts of coastal growth. As a Department of Commerce agency, NOAA is also charged with encouraging sustainable development where appropriate. These objectives of the SAMP program correspond closely with several key goals outlined in the Northampton County Comprehensive Plan, including the preservation of migratory bird habitat (Northampton County Joint Planning Commission 1990).

County, state, federal and non-governmental partners quickly joined together to create and support the SAMP strategy. Wildlife habitat, on the land and in the water, was identified as a valuable and threatened coastal resource and selected for protection and enhancement under the SAMP. Specifically, migratory birds, fin fish, and shellfish became the central themes for conservation and sustainable economic development in the county.

Although this discussion is concerned with landbird stopover habitat conservation, management plans become stronger and generate wider support if policies address more than a single issue. In the case of the Northampton SAMP, the distribution and health of coastal vegetation unites both water quality and stopover opportunities for landbird migrants, as well as rare plants and natural communities. These diverse yet related elements allow for a broader justification for preserving coastal vegetation (Northampton County Board of Supervisors and The Sustainable Development Task Force 1994).

The SAMP strategy set forth four main policy objectives for habitat protection. The first is to control the cumulative and secondary impacts of coastal growth and development by maintaining maximum vegetative cover on land throughout the county. The second is to steer development away from sensitive habitat and groundwater recharge areas. The third objective is to protect water quality, particularly in important fin- and shellfish nursery grounds and aquaculture growout areas. The fourth is to increase public access in appropriate times and places and promote nature tourism. This last objective recognizes that natural resources must be used and enjoyed to be valued, and seeks to insure that the pressure to do so is given a positive, non-damaging outlet. The goal for each objective is to take proactive steps rather than react after conflicts and problems have evolved. In developing and implementing the SAMP, Northampton County seeks to stimulate the local economy and protect sensitive resources at the same time. The task of creating these policies began with two very basic questions—one scientific, the other political. The questions are simply: (1) What *should* be protected? and (2) What *can* be protected?

Groundwork for the answer to the first question was laid in 1991 with a regional study of fall migrant landbird distribution on the Cape May and Delmarva peninsulas. Surveys conducted in New Jersey, Delaware, Maryland, and Virginia established that migrants concentrate within 1.5 km of the coast and that coastal areas on the Delaware and Chesapeake Bays support higher numbers of migrants than those on the Atlantic side of both peninsulas (McCann et al. 1993).

To determine in greater detail what should be protected within Northampton County, the Virginia Department of Conservation and Recreation's Division of Natural Heritage and the Virginia Department of Game and Inland Fisheries' Nongame and Endangered Wildlife Program began a cooperative research project in the fall of 1992. The focus of the study was to define the distribution of migrants in terms of geographic, landscape, and habitat factors. As the investigators for this project, we established a nested design that allowed data collection at several levels simultaneously: the geographic level, the landscape level, and two dimensions within forest patches (distance from edge and vegetational strata). For details of this study see Watts and Mabey (1994). Data were collected during the fall migration periods in 1992 and 1993. Based on nearly 15,000 point counts conducted at almost 200 points throughout the county, Watts and Mabey (1994) estimate that between 6–7 million forest/scrub-dependent migrants pass through Northampton County every year. Further significant results were found at two levels (geographic and habitat) directly relevant to policy development. As a group, long-distance migrants are concentrated within the southernmost 10 km of the peninsula and are more abundant within the bayside coastal forest than seaside coastal forest. Short-distance migrants display a somewhat more even geographic distribution whereas residents are least abundant within 10 km of the peninsula tip. At the habitat level, Watts and Mabey (1994) found that forest patch size had no effect on the distribution of birds within this landscape. Within forests, however, many species of migrants exhibited strong associations with high density understory and subcanopy vegetation.

Based on the these results, the SAMP partners developed a conservation ideal for stopover habitat on the lower Delmarva. This ideal includes three parts: (1) a "no-new-development" overlay zone to cover the lower 10 km of the peninsula and extending up the peninsula within 1 km of the bay coast; (2) maintenance of vegetation throughout the rest of the county at 60–70% of its current coverage, allowing timbering if the forest regeneration that follows is natural; and (3) creation of an incentive program for habitat restoration that would encourage landowners to reforest and plant native shrubs in "dead space." The ideal would serve as a guidepost for protecting the local habitat elements associated with high densities of migrating birds and could be further improved in light of any new research findings.

While our research began to answer to the question of what *should* be protected, the County faced the simultaneous political question of what *could* be protected. From the beginning, the SAMP partners were aware that to achieve even a fraction of the conservation ideal, community support must be cultivated; the citizens would have to care about migrants. The first task in this arena was to create a vehicle for generating that support and producing tangible economic benefits based on birds. The idea for a birding festival was developed under the SAMP as a positive demonstration of natural resource-compatible economic activity. The initiative for the birding festival originally came from the Virginia Coastal Resources Management Program, but the drive and action came from the community. On the weekend of October 9–10, 1993, nearly 1,000 people attended the First Annual Eastern Shore Birding Festival (ESBF) and spent over $36,000 in Northampton County, contributing an estimated $52,300 to local economy (Chazal 1993). Approximately 30% of the attendees were local citizens from Northampton or Accomack County, thus indicating that locals were interested in learning about migratory birds. The festival was linked to Birdlife International's first World BirdWatch, underscoring the international significance of Northampton County's natural habitats. The success of the ESBF is best illustrated by its continued existence and support within the County.

Many other means have been employed for generating community support for migratory birds. SAMP partners have used the local media as a platform for explaining the problems facing neotropical migrants, the migrant-habitat research project, and the concepts of sustainable development. Acting on the premise that every birdwatcher can be an emissary for migrant conservation, SAMP partners have taken the time to talk to anyone who would listen about migratory birds and why they are important, giving formal and informal presentations to local school children, birding clubs, and service clubs, among others. Birdwatchers were asked to do the same and encouraged to display their binoculars wherever they spent money. Perhaps most importantly, the researchers and coastal zone program personnel have been an active and nearly constant presence in the county offices. This has had the threefold benefit of reinforcing the partnerships, ensuring the County that they have dedicated state-level support, and providing the birds a place in policy decisions.

Three major road-blocks have stood between the conservation ideal and implementation. First, politics and science operate on different schedules. This means, in effect, that political decisions are often made without full scientific support. The original SAMP strategy indicated that the County would pursue changes in the zoning code to protect migratory bird habitat. Coincidentally, the zoning commission began revising the zoning code in late 1992. The SAMP interests were introduced into this process somewhat prematurely. Standards for habitat protection were based on the results of one year of research and only a preliminary analysis of the two years' combined data. However, we assumed that revisions later on would be easier than starting the whole process over after the research was complete. For us, it was an uncomfortable but necessary compromise in favor of the political time line.

The second block was that existing conditions imposed strict limitations on the proposed conservation ideal. The suggested "no-new-development" overlay zone was an impossibility. The bayside and lower peninsula are facing the greatest development threats and many large waterfront lots are already sub-divided. Although the houses are yet to be built, prohibiting development in this area could open the door to property rights/takings lawsuits. Tax relief or other incentive programs that theoretically and intuitively would enhance private landowner involvement are currently beyond the capacity of the county budget. Additionally, Virginia state law prevents local regulations from being stricter than state regulations and many creative land-use tools used in other states, including TDRs, are not allowed under state law.

The last barrier to implementing the suggested protection measures involved trade-offs between enforceability and complexity. The more complex policies, regulations, or standards become, the more time and money that are required for enforcement. For example, overlay zones specifying prioritized levels of habitat value and corresponding levels of protection may have been the legally safest option for Northampton County. However, implementing such policy would entail tracking different standards for different areas. This in turn would demand valuable staff time. To the county's credit, they have recognized the limits of the personnel resources available for enforcement and have rejected meaningless paper policy that could never have been properly implemented.

Discussions between planners, biologists, administrators, and citizens involved in the SAMP have been directed toward necessary compromise. At this time, the proposed zoning code for Northampton County includes cluster development zones that apply to new subdivisions only. The purpose of this regulation is to control sprawl, direct development to existing villages and towns, and preserve existing land-use (i.e., agriculture and wooded habitats). The proposed code also contains a new design standard section that applies to existing and new subdivisions, individual lot owners, and new commercial developments. This section limits the percent and location of forest or shrub cover that can be removed from each lot; specifies standards for replacing vegetation if it must be removed beyond set limits; includes a list of native trees and shrubs, highlighting those that are particularly beneficial to wildlife/migratory birds; and prevents landowners from timbering wooded lots to avoid these standards. There is a small but secure victory for stopover habitat in a Memorandum of Understanding (MOU) with the regional power company to manage power line rights-of-way for the maintenance of dense and low native vegetation. As a voluntary conservation approach, the SAMP sponsored the preparation of a landowners' guide to migratory bird habitat management that is distributed through the county planning office. The guide is designed to educate landowners about migrant ecology and habitat needs, as well as to provide a summary of results from the local migrant-habitat research project (Watts and Mabey 1994). It emphasizes the need to preserve dense understory vegetation by minimizing removal of existing vegetation and replacing trees and shrubs to maintain an average vegetation density. Vegetation density, removal and replacement standards recommended in the guide are specific and reflect natural densities in forest patches heavily used by migrants (20 canopy trees, 30 understory trees, and 30 shrubs per acre after development).

The final resolution remains open. The process of developing and implementing local policies is often slow. In this case, local elections changed the composition of the Board of Supervisors and a new set of officials had to be introduced to the history and rationale of the entire SAMP process. The new land-use policies contained within the zoning code have been presented in public hearings and await action from the Board of Supervisors. It remains to be seen if the Northampton community will act on their knowledge of the international significance of the county's stopover habitat and their aesthetic and economic appreciation of migratory birds. However, the success of the Northampton County SAMP project extends beyond the policy itself; it has helped shape the community's evolving attitude toward migratory bird conservation and other natural resources. The act of conducting research and sharing the results with local citizens has gone a long way toward expanding the local possibilities for conservation of migrant habitat. Ultimately, the citizens of Northampton County will have the opportunity to decide the fortune of thousands of acres migratory bird stopover habitat. If conservation is their choice rather than an external imposition, it will have a greater chance of surviving the inevitable land-use conflicts facing rural communities in need of economic development.

CONCLUSION

The critical interplay of unique details prohibits the experience of Northampton County from functioning as a replicable model. Nonetheless, the story illustrates several important points. Science can inform policy decisions and conservation planning efforts and provide legal insurance to local governments wary of property rights lawsuits and land-use conflicts. Scientists can educate a community and provide necessary information for reasoned decisions. Indeed, for conservation efforts to work in a comprehensive fashion, it is critical that scientists work toward breaking down the barriers of language and approach that often separates us from planning and policy professionals.

It is unreasonable to expect most local communities to voluntarily consider habitat essential for migrants or other wildlife in their landuse planning. However, from our experience in Northampton County, we suggest this is a matter of lack of information rather than interest. By sharing our knowledge and concerns with the public, policy makers, and planners, we can contribute locally to the goal of migratory bird conservation. The reward for this effort will be the protection of stopover habitat beyond the reach of state and federal regulation, ranging from a few acres of voluntarily conserved habitat to thousands of acres protected within a well-planned community.

ACKNOWLEDGMENTS

The Northampton County SAMP project is the collective goal and responsibility of many dedicated people. We wish to recognize the invaluable contributions of T. Hayes (SAMP Coordinator), M. Burgard and J. Humphrey (County Planning and Policy Development), and L. McKay (Virginia Coastal Resources Management Program Coastal Projects Coordinator). The project was funded in part by the Virginia Department of Environmental Quality's Coastal Resources Management Program under grant #NA37O20360-01 of the National Oceanic and Atmospheric Administration's Office of Ocean and Coastal Resources Management. Support has also been provided by Virginia Department Conservation and Recreation, Virginia Department of Game and Inland Fisheries, The Center for Conservation Biology, College of William and Mary, and Northampton County and its citizens. This paper has benefited from the comments of P. Kerlinger, T. Litwin, L. McKay, C. Smith, and the University of Southern Mississippi Migratory Bird Group.

Studies in Avian Biology No. 20:109–114, 2000.

ON THE IMPORTANCE OF *EN ROUTE* PERIODS TO THE CONSERVATION OF MIGRATORY LANDBIRDS

RICHARD L. HUTTO

Abstract. In the annual cycle of a migratory landbird, *en route* periods provide unique challenges and selective pressures. The importance of these periods is not yet reflected in the amount of attention this period gets from either research biologists or conservation organizations. Several issues suggest that these annual periods will be important in any program to conserve migratory landbirds: (1) the routes birds take during migration are relatively restricted, implying that some geographic locations may be more critical than others; (2) the habitats migrants use are also relatively restricted and differ between migratory periods, implying that managers will need to know such details to successfully manage for any one species; (3) the migratory periods probably act to limit populations at times, and may be the critical period contributing to long-term declines in some species; and (4) these periods are the source of stories that capture the imagination of humans, implying that basic research during these periods may contribute substantially toward the development of a conservation ethic.

Key Words: conservation, *en route* ecology, habitat selection, neotropical migrants, population regulation, stopover ecology.

In terms of current efforts toward the conservation of migratory landbirds, I propose that the *en route* periods are worthy of considerably more attention than they currently get from research biologists and conservationists. My reasoning involves a discussion of four topics: (1) patterns of geographic distribution of birds while *en route,* (2) patterns of habitat use while *en route,* (3) a possible relationship between changing conditions at stopover sites and population trends, and (4) the way the public at large is captivated by questions and answers associated with *en route* ecology.

GEOGRAPHIC DISTRIBUTION OF *EN ROUTE* MIGRANTS

Geographically speaking, stopover opportunities would appear to be less restricted for landbirds, which can refuel above or within a wide variety of vegetation types, than for other species such as shorebirds, which tend to concentrate in relatively small areas for brief periods of time while *en route* (Myers et al. 1987). Even European migrants that cross the Sahara Desert appear to migrate in broad fronts across the entire expanse of desert rather than within narrowly restricted travel routes (Biebach 1990).

Nonetheless, even though most migratory landbird species can be found over broad fronts during passage, they are not equally abundant in all locations. Moreover, areas of concentrated movement may change from spring to fall, as evidenced by (1) site-specific capture or detection rates, which differ more between migratory seasons than expected due to annual recruitment of young or due to annual mortality (Lavee and Safriel 1989; Winker et al. 1992b,c; Rappole and Ramos 1994, Finch and Yong *this volume*); and by (2) the innovative use of data from museum specimen records, which reveal, for example, that several western hummingbird species migrate up the Pacific coast and back down the Rocky Mountain chain (Phillips 1975; Fig. 1), or that male Hammond's Flycatchers (*Empidonax hammondii*) come up the Pacific coast and females take a more direct inland route later (Johnson 1965). In other words, just because a bird species has been sighted most everywhere at some time or another during migration and, therefore, occurs on spring and fall check-lists across the United States, it does not mean all areas are equally important to the species.

PATTERNS OF HABITAT USE WHILE *EN ROUTE*

Within a fairly restricted geographic location, there are demonstrated nonrandom patterns of habitat use for virtually any landbird species that has ever been studied during either spring or fall migration (e.g., Parnell 1969, Bairlein 1983, Hutto 1985a, Moore et al. 1990, Weisbrod et al. 1993, Winker 1995c, Yong and Finch 1997a, Petit *this volume*). Patterns of habitat use also vary significantly among species (Bairlein 1983, Hutto 1985a, Moore et al. 1995); habitats that receive relatively heavy use by one species are not necessarily the same ones that receive relatively heavy use by other species (Fig. 2). Moreover, because patterns of habitat use while *en route* may differ from patterns of habitat use during the nonmigratory breeding or wintering periods for any given species (Faaborg et al. 1996, Parrish *this volume*), we cannot determine its *en route* needs on the basis of habitat use patterns during one or the other nonmigratory period. As an extreme example, most western species that breed in association with high-elevation and high-latitude conifer forest habitats

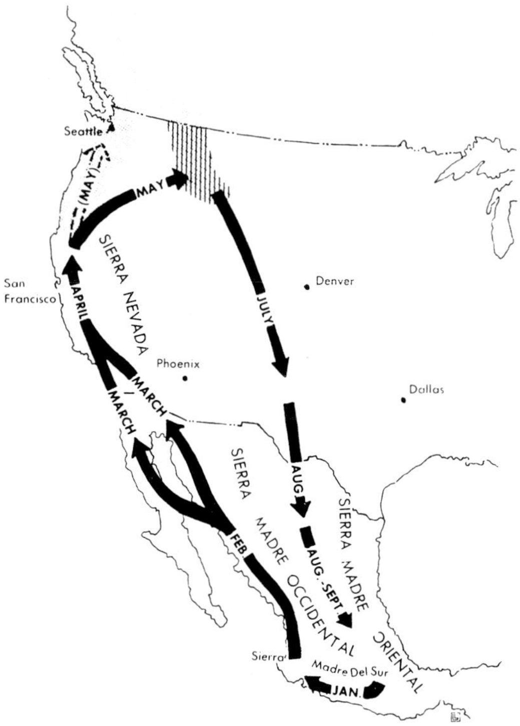

FIGURE 1. The migratory route of the Rufous Hummingbird is not only relatively restricted during a given migratory period, but differs between seasons as well. This pattern was uncovered by plotting the locations of museum specimens that were collected during a given month. Figure taken from Phillips (1975).

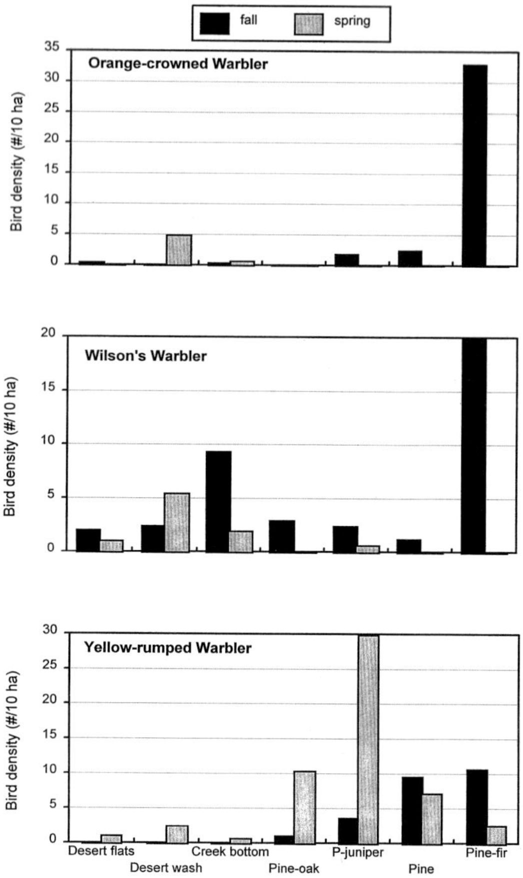

FIGURE 2. Several examples of patterns of habitat use in the Chiricahua Mountains of southern Arizona during both the spring and fall migratory periods. Note that the patterns differ among species, and the patterns differ between seasons for any one species. Data taken from Hutto (1985a).

can be found in good numbers in the Sonoran Desert in spring (pers. observation). Indeed, it is an amazing spectacle to see species like Townsend's (*Dendroica townsendi*) and Hermit (*Dendroica occidentalis*) warblers, which nest high in mature conifer trees, foraging a meter off the ground in creosote bushes in spring!

In addition to nonrandom use of available habitat types, there is also evidence that some species use only those patches that exceed some minimum size, as Cox (1988) has shown with patterns in the springtime use of maritime hammocks in Florida by Black-and-white Warbler (*Mniotilta varia*), Ovenbird (*Seiurus aurocapillus*), Northern Parula (*Parula americana*), and Summer Tanager (*Piranga rubra*), and as Martin (1980, 1981) has shown for shelterbelts of different sizes in South Dakota. There is also evidence that the particular configuration of habitats in the broader landscape may influence the probability that a given patch is used (Simons et al. *this volume*).

En route patterns are generally consistent from year to year (Fig. 3; see also Bairlein 1992a, Winker et al. 1992a), although the pattern during spring passage may differ significantly from the pattern during fall passage (Fig. 2; see also Balda et al. 1975, Winker et al. 1992c, Weisbrod et al. 1993). For a given location, patterns of habitat use may even change with time of day (Moore et al. 1990). We need more information about the patterns of habitat use by migrants during migration. In short, " . . . we do not know what types of habitat are most important, where they occur, and how their distribution and abundance are changing as a result of development and land conversion" (Moore and Simons 1992a).

That there are nonrandom patterns of habitat use, and that the patterns are consistent from year to year indicate that habitats are differentially useful, and that the birds are not simply

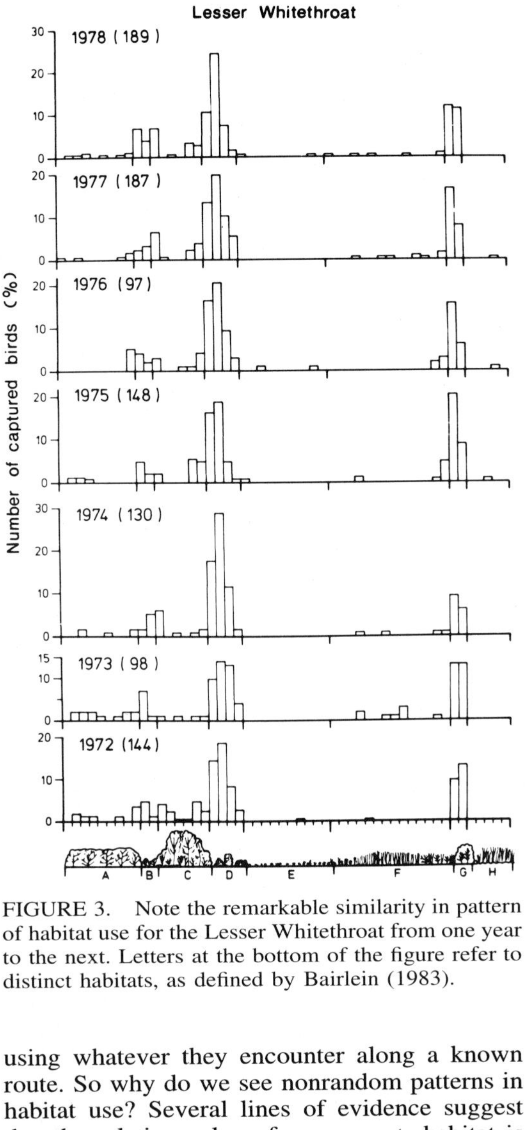

FIGURE 3. Note the remarkable similarity in pattern of habitat use for the Lesser Whitethroat from one year to the next. Letters at the bottom of the figure refer to distinct habitats, as defined by Bairlein (1983).

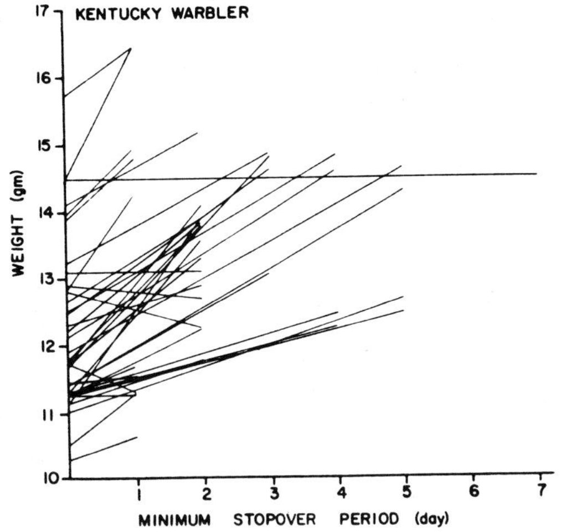

FIGURE 4. Changes in body mass from initial capture to final capture for many individual Kentucky Warblers illustrates the general phenomenon that birds tend to gain mass before departing from a stopover site. Figure taken from Moore and Kerlinger (1987).

using whatever they encounter along a known route. So why do we see nonrandom patterns in habitat use? Several lines of evidence suggest that the relative value of an *en route* habitat is most closely related to the rate at which food can be acquired. Migratory landbirds nearly always gain mass at stopover sites (Winker et al. 1992b, Moore and Kerlinger 1987, Moore et al. 1993, Moore et al. 1995, Winker 1995a, Morris et al. 1996). Migrants are also known to reorient toward areas of greater food availability after landing in food-poor sites (Lindström and Alerstam 1986, Wiedner et al. 1992), and are known to orient differently depending on energetic condition (Sandberg and Moore 1996). Shifts in patterns of habitat use from one migratory period to the other are also apparently in response to shifts in relative availability of food both between (Balda et al. 1975; Hutto 1985a,b), and within (Laursen 1978; Bairlein 1983, 1992a) seasons. Perhaps the most convincing evidence that the primary value of a stopover site is related to the rate at which a bird can gain mass is that fat birds do not remain in a site as long as lean birds (Dolnik and Blyumental 1967, Yong and Moore 1997), and that most birds leave only after gaining mass, as illustrated by Moore and Kerlinger's (1987) data on Kentucky Warblers (*Oporornis formosis*) that stopover in southwest Louisiana after their trans-Gulf flight in spring (Fig. 4). In addition, Carpenter et al. (1983) have shown that Rufous Hummingbirds (*Selasphorus rufus*) adjust territory size on a daily basis in a way that maximizes the rate of weight gain per day, and Russell et al. (1994) provide evidence that survival of those hummingbirds is related to habitat (food) quality. When results from these studies are taken together, there is ample evidence to suggest that food acquisition rate is of primary importance in explaining nonrandom patterns in habitat use during migration (Alerstam and Lindström 1990), although there may be predation or other constraints that make the suitability of habitats somewhat different from that which would be predicted on the basis of food availability alone (Lindström 1989, 1990b).

Even though food acquisition may contribute disproportionately to the suitability of a site, ranking the suitability of habitats or sites (as may be desired to help set conservation priorities) is probably not as easy as measuring and comparing average fat loads, feeding rates, or stopover durations among sites because (1) fat birds may use different habitats from lean birds (Biebach et al. 1986); (2) feeding rate depends on current body condition (Rappole and Warner 1976, Yong and Moore 1993, Moore 1994), age (Woodrey *this volume*), and competitive milieu (Moore and Yong 1991); and (3) duration of stay is related to fat level (Cherry 1982, Biebach 1985, Moore and Kerlinger 1987, Lavee and Safriel 1989, Loria and Moore 1990, Kuenzi et al. 1991), age (Ellegren 1991; Morris et al. 1994, 1996; Woodrey and Moore 1997, Woodrey *this volume*), sex (Morris et al. 1994, Otahal 1995), and possibly even the status of a migrant's biological clock (Safriel and Lavee 1988). Thus, significant differences in the "average" age, sex, body condition, or time since arrival among sites will make a meaningful comparison of average fat loads, feeding rates or stopover durations difficult at best. Ranking the relative suitability of habitats or sites using such information will probably necessitate labor intensive capturing, marking, and re-capturing of birds so that confounding variables such as age, sex, body condition, and time since arrival can be factored out before comparing fat loads, feeding rates, or stopover durations among sites.

Even though we are learning what to measure in order to rank habitats according to their suitability, we must also be careful not to get too carried away with thinking we can accurately rank suitabilities of habitats or specific locations for at least two other reasons: (1) different "strategies" of weight gain and load may exist for different-sized birds (Yong and Moore 1994) or for birds of different populations that migrate as little as 150 miles apart (Karlsson et al. 1988); and (2) both intrinsic and extrinsic factors (sensu Hutto 1985b) unrelated to food acquisition may contribute to a location's suitability as a stopover site, including low predation rates or its geographic position relative to a migratory route that is restricted for physiographic or climatic reasons. Thus, in terms of food acquisition, a habitat or specific location might rank low in comparison with all others for which there are data, but it may still be the best thing going in certain geographic locations.

THE IMPORTANCE OF MIGRATORY PERIODS TO POPULATION REGULATION

A key question critical to conservation efforts is: How likely is it that the migratory period contributes to long-term population trends? Even without relevant data, most authors of research proposals and other material involving *en route* ecology generally claim that these periods are critical. Are they? Don't migrants seem to use whatever they come across, occur most everywhere, and do well with whatever exists? Or, to echo the sentiments expressed by most participants in the first Smithsonian conference on northern migrants in the tropics (Buechner and Buechner 1970): Aren't migrants flexible enough to withstand any change we throw at them?

How do we assess the importance of any season in terms of its contribution to long-term population trends? Over a short period of time, it seems clear that year-to-year trends are controlled by events in different seasons in different years (a summer insect outbreak in one year, a bad winter in another, a springtime storm in another), and that all seasons are important in that sense (Sherry and Holmes 1993). But what about long-term trends? We need to recognize that long-term trends are a separate issue from short-term, year-to-year population fluctuations. While short-term trends are equally likely to be consequences of events in summer, winter, spring, or fall, depending on the year, longer-term population trends can, at the same time, be controlled by events in a single season.

We could evaluate the importance of migratory periods on theoretical grounds, and Sherry and Holmes (1992, 1993, 1995) provide recent reviews of this approach, which generally supports the idea that migratory periods are indeed important. Using a more empirical approach, Svensson (1978) showed that spring numbers predict breeding numbers later on, and suggested that the spring migratory period can be an important contributor to population trends. The problem with the latter approach, however, is if a population is undergoing a long-term change, numbers in *any* season will be correlated with numbers at some later point in time. Ideally, we need population data from just before and just after the season of interest to see if the direction and magnitude of change affects the number seen at some subsequent point in time, year after year (Owen and Black 1991). Unfortunately, it will be difficult to amass such data for at least two reasons. First, it is difficult to define and follow a single population unit year round. For example, it would not be easy to make sense of data from a well defined Montana breeding group that then splits into winter populations scattered from California through Oaxaca. Second, the time period needed for such study is generally beyond the scope of most individual research programs.

A more feasible alternative is to *project* inevitable trends by coupling results from intensive short-term studies of habitat associations with estimates of land-use changes and the anticipated distribution and abundance of habitats into the future. Even here, population projections will be difficult because (1) a habitat that is not used in one place may be important in another, so the "value" of a habitat may vary from place to place, and will require the development of regional models of habitat use; and (2) human-altered habitats have greater potential than naturally occurring habitats to act as ecological traps (sensu Gates and Gysel 1978) because human habitat alteration often uncouples normally co-occurring habitat elements such that proximately and ultimately important factors are no longer linked. Thus, a bird can end up being attracted to a site with appropriate proximate cues but inappropriate ultimately important conditions. This, in turn, means that abundance data alone may not reflect the relative suitabilities of habitats created through a mixture of both natural and unnatural processes. We need data on feeding rates and other characteristics of marked birds, as described earlier, but the collection of such data is labor intensive relative to the collection of data on occurrence among habitats.

So, several lines of evidence suggest that migratory periods are important to the conservation of migrants, but exactly how the management of lands used by migrants *en route* will affect population trends is going to be hard to determine.

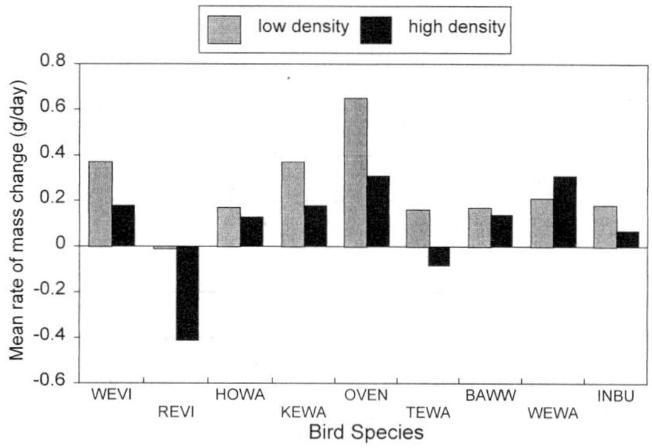

FIGURE 5. The rate of mass gain for a variety of species is inversely related to the density of other migrants in the area, implying that interspecific competition may be important selective pressure during the migratory period. Bird species include White-eyed Vireo (WEVI), Red-eyed Vireo (REVI), Hooded Warbler (HOWA), Kentucky Warbler (KEWA), Ovenbird (OVEN), Tennessee Warbler (TEWA), Black-and-white Warbler (BAWW), Worm-eating Warbler (WEWA), and Indigo Bunting (INBU). Data taken from Moore and Yong (1991).

STORY-TELLING POWER ASSOCIATED WITH *EN ROUTE* PERIOD

One last consideration suggests that no matter what role migratory periods play in terms of the regulation of migratory bird populations, stopover biology and the *en route* periods will still be a key to the conservation of migrants. That consideration involves the story-telling power of this part of the annual cycle of birds, which is immeasurable. For example,

•Why should birds that cross the Sahara stop while *en route*? Because headwinds develop at night, most passerines would not meet the energetic costs of flying both day and night if they stayed at the typical daytime flight altitude of 3000 m. They could fly at a lower altitude during the night, where head winds are less likely; however, they would then move into a warmer zone where the air temperature would cause water loss to become a problem. Thus, the only option is to stop at night (Biebach 1990, Leberg et al. 1996)!

•The same hummingbird that was banded by Elly Jones near Swan Lake, Montana, was caught 10 days later and about 1000 miles south at the Rocky Mountain Biological Laboratory in Colorado by William Calder, the biologist who taught Elly Jones to band hummingbirds!

•While *en route*, many landbird species travel in pairs (Greenberg and Gradwohl 1980) or small groups (Moore 1990b), and may even copulate (Quay 1989, Moore and McDonald 1993) long before they arrive on the "breeding" grounds!

•Recent work suggests that some landbird species may stay for long periods at some "stopover" sites to molt, before continuing farther south to "winter" (Hedenström et al. 1993).

•Studies of small migratory birds while *en route* provide some remarkably clear demonstrations of resource depletion and competition in bird communities. For example, Pied Flycatchers deplete resources, as evidenced by declining capture rates with time in same area and by increasing feeding rates with time since last visit to a given tree (Bibby and Green 1980). That there may be interspecific effects from such food depletion is also suggested because feeding rates are greater on days when relatively few competitors are present (Fig. 5; Moore and Yong 1991), and diets overlap less when there are more potential competitors present (Laursen 1978).

I hope these examples serve to illustrate that fascinating stories emerge from studies of basic biology of landbirds during migration. My point is that both research biologists and conservation-oriented funding agencies may be putting too much emphasis on conducting or funding conservation projects that involve "high priority" species, and too little attention on other measures of research attractiveness. Our efforts to determine priorities for research that will help conserve migrants will be misdirected if we fail to consider the story-telling power of proposed research, whether the nature of the work relates directly to saving a high priority species or not! Conservation success is not measured by whether we use limited resources in the right way to save one species; rather, success is measured by how much of the natural world (how many species in total) we can retain. And that is going to be directly related to people's collective attitude about conservation, which may itself be affected less by whether we save a priority species than by whether people have developed a connection between themselves and nature (see also Mabey and Watts *this volume*). That connection to nature, in turn, comes from story telling. Thus, maybe the spending priority of conservation organizations should shift toward support of solid biological research by people who have an eye toward what makes an interesting study, period. Those efforts are most likely to enhance our understanding of the biology of organisms, produce good stories, and change people's attitudes about conservation. Because of the fascinating questions that become apparent to anyone who has read about stopover biology, that area of research should play an integral part in any conservation program that considers an accepting public to be an essential ingredient to success.

Studies in Avian Biology No. 20:115–133, 2000.

LITERATURE CITED

ABLE, K. P. 1972. Fall migration in coastal Louisiana and the evolution of migration patterns in the Gulf region. Wilson Bulletin 84:231–242.

ABLE, K. P. 1977. The orientation of passerine nocturnal migrants following offshore drift. Auk 94:320–330.

ABORN, D. A. 1994. Correlation between raptor and songbird numbers at a migratory stopover site. Wilson Bulletin 106:150–154.

ABORN, D. A., AND F. R. MOORE. 1997. Pattern of movement by Summer Tanagers (*Piranga rubra*) during migratory stopover: a telemetry study. Behaviour 134:1077–1100.

ABRAMSKY, Z., AND U. SAFRIEL. 1980. Seasonal patterns in a Mediterranean bird community composed of transient, wintering and resident passerines. Ornis Scandinavica 11:201–216.

AGARD, K. 1995. Migrant songbirds along Lake Ontario's Shore. The Nature Conservancy, Rochester, NY.

ALATALO, R. V. 1981. Habitat selection of forest birds in the seasonal environment of Finland. Annales Zoologici Fennici 19:103–114.

ALATALO, R. V., L. GUSTAFSSON, AND A. LUNDBERG. 1984. Why do young passerine birds have shorter wings than older birds? Ibis 126:410–415.

ALDRICH, J. W., AND C. S. ROBBINS. 1970. Changing abundance of migratory birds in North America. Pp. 17–26 *in* H. K. Buechner and J. H. Buechner (editors). The avifauna of northern Latin America. Smithsonian Contributions to Zoology 26. Smithsonian Institution, Washington, D.C.

ALERSTAM, T. 1978. Reoriented bird migration in coastal areas: dispersal to suitable resting grounds? Oikos 30:405–408.

ALERSTAM, T. 1990. Ecological causes and consequences of bird orientation. Experientia 46:405–415.

ALERSTAM, T. 1993. Bird migration. Cambridge University Press, Cambridge, UK.

ALERSTAM, T., AND P. H. ENCKELL. 1979. Unpredictable habitats and the evolution of bird migration. Oikos 33:228–232.

ALERSTAM, T., AND G. HÖGSTEDT. 1982. Bird migration and reproduction in relation to habitats for survival and breeding. Ornis Scandinavica 13:25–37.

ALERSTAM, T., AND Å. LINDSTRÖM. 1990. Optimal bird migration: the relative importance of time, energy and safety. Pp. 331–351 *in* E. Gwinner (editor). Bird migration: the physiology and ecophysiology. Springer-Verlag, Berlin, FRG.

ALLEN, D. L. 1967. The life of prairies and plains. McGraw-Hill, New York, NY.

AMERICAN RIVERS. 1993. The nation's ten most endangered rivers and fifteen most threatened rivers for 1993. American Rivers, Washington, D.C.

ARMISTEAD, H. T. 1993. The autumn migration: middle Atlantic coast region. American Birds 47:75–79.

ASKINS, R. A., J. F. LYNCH, AND R. S. GREENBERG. 1990. Population declines in migratory birds in eastern North America. Current Ornithology 7:1–57.

ASKINS, R. A., M. J. PHILBRICK, AND D. SUGENO. 1987. Relationship between the regional abundance of forest and composition of forest bird communities. Biological Conservation 39:129–152.

BAILLIE, S. R., AND W. J. PEACH. 1992. Population limitation in palearctic-African migrant passerines. Ibis 134 (suppl. 1):120–132.

BAIRD, J. W. 1980. The selection and use of fruit by birds in an eastern forest. Wilson Bulletin 92:63–73.

BAIRLEIN, F. 1983. Habitat selection and associations of species in European passerine birds during southward, post-breeding migrations. Ornis Scandinavica 14:239–245.

BAIRLEIN, F. 1985a. Body weights and fat deposition of palearctic passerine migrants in the central Sahara. Oecologia 66:141–146.

BAIRLEIN, F. 1985b. Efficiency of food utilization during fat deposition in the long-distance migratory garden warbler, *Sylvia borin*. Oecologia 68:118–125.

BAIRLEIN, F. 1987a. Nutritional requirements for maintenance of body weight and fat deposition in the long-distance migratory Garden Warbler, *Sylvia borin* (Boddaert). Comparative Biochemistry and Physiology 86A:337–347.

BAIRLEIN, F. 1987b. The migratory strategy of the Garden Warbler: a survey of field and laboratory data. Ringing and Migration 8:59–72.

BAIRLEIN, F. 1990. Nutrition and food selection in migratory birds. Pp. 198–213 *in* E. Gwinner (editor). Bird migration: the physiology and ecophysiology. Springer-Verlag, Berlin, FRG.

BAIRLEIN, F. 1991. Body mass of Garden Warblers (*Sylvia borin*) on migration: a review of field data. Die Vogelwarte 36:48–61.

BAIRLEIN, F. 1992a. Morphology-habitat relationships in migrating songbirds. Pp. 356–369 *in* J. M. Hagan, III, and D. W. Johnston (editors). Ecology and conservation of neotropical migrant landbirds. Smithsonian Institution Press, Washington, D.C.

BAIRLEIN, F. 1992b. Recent prospects on trans-Sahara migration of songbirds. Ibis 134 (suppl. 1):41–46.

BAKER, R. R. 1978. The evolutionary ecology of animal migration. Holmes and Meier, New York, NY.

BALDA, R. P., B. C. MCKNIGHT, AND C. D. JOHNSON. 1975. Flammulated Owl migration in the southwestern United States. Wilson Bulletin 87:520–533.

BANCROFT, G. T., W. HOFFMAN, R. J. SAWICKI, AND J. C. OGDEN. 1992. The importance of water conservation areas in the Everglades to the endangered Wood Stork (*Mycteria americana*). Conservation Biology 6:392–398.

BARLOW, J. C. 1980. Patterns of ecological interactions among migrant and resident vireos on the wintering grounds. Pp. 79–107 *in* A. Keast and E. S. Morton (editors). Migrant birds in the Neotropics: ecology, behavior, distribution, and conservation. Smithsonian Institution Press, Washington, D.C.

BARROW, W. C., JR. 1990. Ecology of small insectivorous birds in a bottomland hardwood forest. Ph.D. dissertation. Louisiana State University, Baton Rouge, LA.

BASCIUTTI, P., O. NEGRA, AND F. SPINA. 1997. Autumn migration strategies of the Sedge Warbler *Acrocephalus schoenbaenus* in northern Italy. Ringing and Migration 18:59–67.

BEAL, K. G., AND H. J. KHAMIS. 1991. A problem in

statistical analysis: simultaneous inference. Condor 93:1023–1025.

BEAN, M. J., AND D. S. WILCOVE. 1997. The private-land problem. Conservation Biology 11:3–5.

BEAVER, D. L. 1988. The response of bird populations to three years of wastewater irrigation on oldfields in Michigan. Jack-Pine Warbler 66:87–102.

BEEDY, E. C. 1981. Bird communities and forest structure in the Sierra Nevada of California. Condor 83: 97–105.

BELL, W. J. 1991. Searching behaviour: the behavioural ecology of finding resources. Chapman and Hall, London, UK.

BELLROSE, F. C. 1976. Ducks, geese, and swans of North America. Wildlife Management Institute, Harrisburg, PA.

BENDER, G. L. (EDITOR). 1982. Reference handbook on the deserts of North America. Greenwood Press, Westport, CT.

BERTHOLD, P. 1975. Migration: control and metabolic physiology. Pp. 77–128 *in* D. S. Farner and J. R. King (editors). Avian biology, vol. V. Academic Press, New York, NY.

BERTHOLD, P. 1976a. Animalische und vegetabilische ernahrung omnivorer Singvogelgarten: Nahrungsbevorzugung, jahresperiodik der nahrungswahl, physiologische und okologische. Journal für Ornithologie 117:145–209.

BERTHOLD, P. 1976b. The control and significance of animal and vegetable nutrition in omnivorous songbirds. Ardea 64:140–154.

BERTHOLD, P. 1984. The endogenous control of bird migration: a survey of the experimental evidence. Bird Study 31:19–27.

BERTHOLD, P. 1988. The control of migration in European warblers. Acta XIX Congressus Internationalis Ornithologici 19:215–249.

BERTHOLD, P. 1990. Genetics of migration. Pp. 269–280 *in* E. Gwinner (editor). Bird migration: the physiology and ecophysiology. Springer-Verlag, Berlin, FRG.

BERTHOLD, P. 1993. Bird migration. Oxford University Press, New York, NY.

BERTHOLD, P., AND A. J. HELBIG. 1992. The genetics of bird migration: stimulus, timing and direction. Ibis 134 (suppl. 1):35–40.

BIBBY, C. F. 1992. Conservation of migrants on their breeding grounds. Ibis 134 (suppl. 1):29–34.

BIBBY, C. F., AND R. E. GREEN. 1980. Foraging behavior of migrant Pied Flycatcher, *Ficedula hypoleuca,* on temporary territories. Journal of Animal Ecology 49:507–521.

BIBBY, C. F., AND R. E. GREEN. 1981. Autumn migration strategies of Reed and Sedge warblers. Ornis Scandinavica 12:1–12.

BIBBY, C. F., AND R. E. GREEN. 1983. Food and fattening of migrating warblers in some French marshlands. Ringing and Migration 4:175–184.

BIBBY, C. F., R. E. GREEN, G. R. M. PEPLER, AND P. A. PEPLER. 1976. Sedge Warbler migration and reed aphids. British Birds 69:384–399.

BIEBACH, H. 1985. Sahara stopover in migratory flycatchers: fat and food affect the time program. Experientia 41:695–697.

BIEBACH, H. 1990. Strategies of trans-Sahara migrants. Pp. 352–367 *in* E. Gwinner (editor). Bird migration: the physiology and ecophysiology. Springer-Verlag, Berlin, FRG.

BIEBACH, H. 1992. Flight-range estimates for small trans-Sahara migrants. Ibis 134 (suppl. 1):47–54.

BIEBACH, H., W. FRIEDRICH, AND G. HEINE. 1986. Interaction of body mass, fat, foraging and stopover period in trans-Sahara migrating passerine birds. Oecologia 69:370–379.

BILCKE, G. 1984. Seasonal changes in habitat use of resident passerines. Ardea 72:95–99.

BILLINGS, F. H. 1909. Some observations on the motile leaves of *Erythrina herbacea.* Gulf Biologic Station Bulletin 13:1.

BINGHAM, B. B., AND B. R. NOON. 1997. Mitigation of habitat "takes": application to habitat conservation planning. Conservation Biology 11:127–139.

BINGMAN, V. P. 1980. Inland morning flight behavior of nocturnal passerine migrants in eastern New York. Auk 97:465–472.

BINGMAN, V. P., K. P. ABLE, AND P. KERLINGER. 1982. Wind drift, compensation, and the use of landmarks by nocturnal bird migrants. Animal Behavior 30:49–53.

BIRKHEAD, T. R. 1988. Behavioral aspects of sperm competition in birds. Advances in the Study of Behavior 18:35–72.

BLAKE, J. G. 1982. Influence of fire and logging on nonbreeding bird communities of ponderosa pine forests. Journal of Wildlife Management 46:404–415.

BLAKE, J. G. 1984. A seasonal analysis of bird communities in southern Nevada. Southwestern Naturalist 29:463–474.

BLAKE, J. G., AND W. G. HOPPES. 1986. Influence of resource abundance on use of tree-fall gaps by birds in an isolated woodlot. Auk 103:328–340.

BLAKE, J. G., AND B. A. LOISELLE. 1991. Variation in resource abundance affects capture rates of birds in three lowland habitats in Costa Rica. Auk 108:114–130.

BLAKE, J. G., AND B. A. LOISELLE. 1992a. Fruits in the diets of neotropical migrant birds in Costa Rica. Biotropica 24:200–210.

BLAKE, J. G., AND B. A. LOISELLE. 1992b. Habitat use by neotropical migrants at La Selva Biological Station and Braulio Carrillo National Park, Costa Rica. Pp. 257–272 *in* J. M. Hagan, III, and D. W. Johnston (editors). Ecology and conservation of neotropical migrant landbirds. Smithsonian Institution Press, Washington, D.C.

BLEM, C. R. 1980. The energetics of migration. Pp. 175–224 *in* S. A. Gauthreaux, Jr. (editor). Animal migration, orientation, and navigation. Academic Press, London, UK.

BLOCK, W. M., AND L. A. BRENNAN. 1993. The habitat concept in ornithology: theory and applications. Current Ornithology 11:35–91.

BLONDEL, J. 1969. Synecologie des passereaux residents et migrateurs dans le midi mediterraneen francais. Centre Regional de Docuentation Pedagogique, Marseille, France.

BOCK, C. E., J. H. BOCK, W. R. KENNEY, AND V. M. HAWTHORNE. 1984. Responses of birds, rodents, and vegetation to livestock exclosure in a semi-desert grassland site. Journal of Range Management 37: 239–242.

BONGAARTS, J. 1994. Population policy options in the developing world. Science 263:771–776.

BOROWICZ, V. A. 1988. Fruit consumption by birds in relation to fat content of pulp. American Midland Naturalist 119:121–127.

BRIGGS, S. A., AND J. H. CRISWELL. 1978. Gradual silencing of spring in Washington: selective reduction of species of birds found in three woodland areas over the past 30 years. Atlantic Naturalist 32:19–26.

BROWN, J. L. 1963. Aggressiveness, dominance and social organization in the Steller's Jay. Condor 65: 460–484.

BROWN, J. L. 1969. Territorial behavior and population regulation in birds. Wilson Bulletin 81:293–329.

BROWN, J. L. 1975. The evolution of behavior. Norton, New York, NY.

BROWN, L. 1985. Grasslands. The Audubon Society Nature Guides. Alfred A. Knopf, New York, NY.

BRUCE, K. A. 1993. Factors affecting the biological invasion of the exotic Chinese tallow tree, *Sapium sebiferum,* in the Gulf coast prairie of Texas. M.S. thesis. University of Houston, Houston, TX.

BRUCE, K. A., G. N. CAMERON, AND P. A. HARCOMBE. 1995. Initiation of a new woodland type on the Texas coastal prairie by the Chinese tallow tree (*Sapium sebiferum* (L.) Roxb.). Bulletin of the Torrey Botanical Club 122(3):215–225.

BRUDERER, B., AND L. JENNI. 1988. Strategies of bird migration in the area of the Alps. Acta XIX Congressus Internationalis Ornithologici 19:2150–2161.

BRUDERER, B., AND L. JENNI. 1990. Migration across the Alps. Pp. 60–77 *in* E. Gwinner (editor). Bird migration: the physiology and ecophysiology. Springer-Verlag, Berlin, FRG.

BUECHNER, H. K., AND J. H. BUECHNER (EDITORS). 1970. The avifauna of northern Latin America. Smithsonian Contributions to Zoology 26. Smithsonian Institution, Washington, D.C.

BULLARD, T. F., AND S. G. WELLS. 1992. Hydrology of the middle Rio Grande from Velarde to Elephant Butte reservoir, New Mexico. U.S. Department of Interior, Fish and Wildlife Service Research Publication 179.

BURGER, J. 1988. Effects of age on foraging in birds. Acta XIX Congressus Internationalis Ornithologici 19:1127–1140.

BUSKIRK, W. H. 1980. Influence of meteorological patterns and trans-Gulf migration on the calendars of latitudinal migrants. Pp. 485–491 *in* A. Keast and E. S. Morton (editors). Migrant birds in the Neotropics: ecology, behavior, distribution, and conservation. Smithsonian Institution Press, Washington, D.C.

BYRNE, J. V., D. O. LEROY, AND C. M. RILEY. 1959. The Chenier Plain and its stratigraphy, southwest Louisiana. Transactions of the Gulf Coast Association Geological Society 9:237–270.

CAROTHERS, S. W., R. R. JOHNSON, AND S. W. AITCHISON. 1974. Population structure and social organization of Southwest riparian birds. American Zoologist 14:97–108.

CARPENTER, F. L., M. A. HIXON, R. W. RUSSELL, D. C. PATON, AND E. J. TEMELES. 1993a. Interference asymmetries among age-sex classes of Rufous Hummingbirds during migratory stopovers. Behavioral Ecology and Sociobiology 33:297–304.

CARPENTER, F. L., M. A. HIXON, E. J. TEMELES, R. W. RUSSELL, AND D. C. PATON. 1993b. Exploitative compensation by subordinate age-sex classes of migrant Rufous Hummingbirds. Behavioral Ecology and Sociobiology 33:305–312.

CARPENTER, F. L., D. C. PATON, AND M. A. HIXON. 1983. Weight gain and adjustment of feeding territory size in migrant hummingbirds. Proceedings of the National Academy of Sciences (USA) 80:7259–7263.

CARROLL, M. S., AND W. G. HENDRIX. 1992. Federally protected rivers: the need for effective local involvement. Journal of the American Planning Association 58:346–352.

CARSON, R. 1962. Silent Spring. Houghton Mifflin, Boston, MA.

CHARNOV, E. R. 1976. Optimal foraging: the marginal value theorem. Theoretical Population Biology 9: 129–136.

CHAZAL, J. 1993. Economic impacts of the 1993 Eastern Shore Birding Festival. Virginia Department of Environmental Quality, Grant No. NA27OZ0312-03, Richmond, VA.

CHEN, C. C. 1996. Foraging behavior and habitat selection of insectivorous migratory songbirds at Gulf coast stopover sites in spring. Ph.D. dissertation. Louisiana State University, Baton Rouge, LA.

CHERRY, J. D. 1982. Fat deposition and length of stopover of migrant White-crowned Sparrows. Auk 99: 725–732.

CHERRY, J. D. 1985. Early autumn movements and prebasic molt of Swainson's Thrushes. Wilson Bulletin 97:368–370.

CHILD, G. L. 1969. A study of nonfat weights in migrating Swainson's Thrushes (*Hylocichla ustulatus*). Auk 86:327–338.

CIMPRICH, D., AND F. R. MOORE. In press. Energetic constraints and predation pressure during stopover. Acta XXI Congressus Internationalis Ornithologici.

CITIZENS FOR A BETTER EASTERN SHORE NEWSLETTER. 1991. December Issue. Eastville, VA.

CLAWSON, R. L. 1982. The status distribution, and habitat preferences of the birds of Missouri. Terrestrial Series No. 11. Missouri Department of Conservation, Jefferson City, MO.

COBLENTZ, B. E. 1990. Exotic organisms: a dilemma for conservation biology. Conservation Biology 4: 261–265.

COCKS, R. S. 1904. Report of the flora in the vicinity of the Gulf Biologic Station. Gulf Biologic Station Bulletin 2:56–64.

COCKS, R. S. 1907. The flora of the Gulf Biologic Station. Gulf Biologic Station Bulletin 1:1–42.

COLEMAN, S. L., AND R. L. MELLGREN. 1994. Neophobia when feeding alone or in flocks in zebra finches, *Taeniopygia guttata.* Animal Behavior 48:903–907.

COLLINS, B. G., AND P. BRIFFA. 1982. Seasonal variation of abundance and foraging of three species of Australian honeyeaters. Australian Wildlife Research 9:557–569.

COLWELL, R. K., AND D. J. FUTUYMA. 1971. On the

measurement of niche breadth and overlap. Ecology 52:567–576.

CONWAY, C. J., W. R. EDDLEMAN, AND K. L. SIMPSON. 1994. Seasonal changes in fatty acid composition of the wood thrush. Condor 96:791–794.

COOKE, W. W. 1915. Bird migration. U.S. Department of Agriculture Bulletin No. 185, Washington, DC.

COOPER, R. J., AND R. C. WHITMORE. 1990. Arthropod sampling methods in ornithology. Studies in Avian Biology 13:29–37.

COX, J. 1988. The influence of forest size on transient and resident bird species occupying maritime hammocks of northeastern Florida. Florida Field Naturalist 16:25–34.

COX, J. 1995. Way stations for migrants. Florida Wildlife (March–April):2–4.

CRAWFORD, C. S., A. C. CULLY, R. LEUTHEUSER, M. S. SIFUENTES, L. H. WHITE, AND J. P. WILBER. 1993. Middle Rio Grande Ecosystem: Bosque biological management plan. U.S. Fish and Wildlife Service, Albuquerque, NM.

CRESSWELL, W. 1994. Age-dependent choice of Redshank (*Tringa totanus*) feeding location: profitability or risk? Journal of Animal Ecology 63:589–600.

CRICK, H. Q. P., AND P. J. JONES (EDITORS). 1992. The ecology and conservation of palearctic-African migrants. Ibis 134 (suppl. 1). Blackwell Scientific Publications, Oxford, UK.

CULLITON, T. J., M. A. WARREN, T. R. GOODSPEED, D. G. REMER, C. M. BLACKWELL, AND J. J. MCDONOUGH, III. 1990. Fifty years of population change along the nation's coasts, 1960–2010. Coastal Trends Series, Report No. 2, NOAA Strategic Assessment Branch, Rockville, MD.

CURIO, E. 1983. Why do young birds reproduce less well? Ibis 125:400–404.

CURIO, E. 1993. Proximate and developmental aspects of anti-predator behavior. Advances in the Study of Behavior 22:135–238.

CURRY-LINDAHL, K. 1981. Bird migration in Africa. Academic Press, New York, NY.

DATE, E. M., H. A. FORD, AND H. F. RECHER. 1991. Frugivorous pigeons, stepping stones, and weeds in northern New South Wales. Pp. 241–245 *in* D. A. Saunders and R. J. Hobbs (editors). Nature conservation 2: the role of corridors. Surrey Beatty and Sons, Chipping Norton, Australia.

DAVIDAR, P., AND E. S. MORTON. 1986. The relationship between fruit crop sizes and fruit removal rates by birds. Ecology 67:262–265.

DAVIS, D. E., AND I. R. SAVIDGE. 1971. Distribution of certain birds in various types of vegetation. Bird-Banding 42:264–268.

DEBUSSCHE, M., J. CORTEZ, AND I. RIMBAULT. 1987. Variation in fleshy fruit composition in the Mediterranean region: the importance of ripening season, life-form, fruit type, and geographical distribution. Oikos 49:244–252.

DELCOURT, H. R., AND P. A. DELCOURT. 1974. Primeval magnolia-holly-beech climax in Louisiana. Ecology 55:638–644.

DESANTE, D. F. 1973. Analysis of the fall occurrences and nocturnal orientations of vagrant warblers (Parulidae) in California. Ph.D. dissertation. Stanford University, Stanford, CA.

DESANTE, D. F. 1983. Annual variability in the abundance of migrant landbirds on southeast Farallon Island, California. Auk 100:826–852.

DESANTE, D. F., AND T. L. GEORGE. 1994. Population trends in the landbirds of western North America. Studies in Avian Biology 15:173–190.

DICK-PEDDIE, W. A. 1993. New Mexico vegetation: past, present and future. University of New Mexico Press, Albuquerque, NM.

DIGIOIA, H. G. 1974. Fall warblers in a suburban woodland. Oriole 39:1–5.

DINGLE, H. 1980. Ecology and evolution of migration. Pp. 1–101 *in* S. A. Gauthreaux Jr. (editor). Animal migration, orientation, and navigation. Academic Press, New York, NY.

DINSMORE, J. J., T. H. KENT, D. KOENIG, P. C. PETERSEN, AND D. M. ROOSA. 1984. Iowa Birds. Iowa State University Press, Ames, IA.

DOLNIK, V. R., AND T. I. BLYUMENTAL. 1967. Autumnal premigratory and migratory periods in the Chaffinch (*Fringilla coelebs coelebs*) and some other temperate-zone passerine birds. Condor 69:435–468.

DRURY, W. H., AND J. A. KEITH. 1962. Radar studies of songbird migration in coastal New England. Ibis 104:449–489.

DUNNING, J.B. 1993. CRC handbook of avian body masses. CRC Press, Boca Raton, FL.

DWYER, L. E., D. D. MURPHY, AND P. R. EHRLICH. 1995. Property rights case law and the challenge to the Endangered Species Act. Conservation Biology 9: 725–741.

EDDINS, M. E., AND D. T. ROGERS. 1992. Autumnal migration of the Gray Catbird through coastal Alabama. Journal of Field Ornithology 63:401–407.

EDWARDS, P. J., R. M. MAY, AND N. R. WEBB. 1994. Large-scale ecology and conservation biology. Blackwell Scientific Publications, Oxford, UK.

EHRLICH, P. R., D. S. DOBKIN, AND D. WHEYE. 1988. The birder's handbook. Simon and Schuster, New York, NY.

EKMAN, J. B., AND C. ASKENMO. 1984. Social rank and habitat use in Willow Tit groups. Animal Behavior 32:508–514.

ELLEGREN, H. 1991. Stopover ecology of autumn migrating Bluethroats *Luscinia s. svecica* in relation to age and sex. Ornis Scandinavica 22:340–348.

ELLIS, L. M. 1995. Bird use of saltcedar and cottonwood vegetation in the middle Rio Grande Valley of New Mexico, U.S.A. Journal of Arid Environments 30:339–349.

ELY, C. A. 1970. Migration of Least and Traill's flycatchers in west-central Kansas. Bird-Banding 41: 198–204.

EMLEN, J. T. 1977. Estimating breeding season bird densities from transect counts. Auk 94:455–468.

ENDICOTT, E. (EDITOR). 1993. Land conservation through public/private partnerships. Island Press, Washington, D.C.

ERDELSON, M. 1984. Bird communities and vegetation structure: I. Correlations and comparisons of simple and diversity indices. Oecologia 61:277–284.

FAABORG, J., A. D. ANDERS, M. E. BALTZ, AND W. K. GRAM. 1996. Non-breeding season considerations for the conservation of migratory birds in the Midwest: post-breeding and wintering periods. Pp. 189–

199 *in* F. R. Thompson, III (editor). Management of midwestern landscapes for the conservation of neotropical migratory birds. USDA Forest Service Gen. Tech. Rep. NC-187. USDA Forest Service North Central Experiment Station, St. Paul, MN.

FAABORG, J., M. BRITTINGHAM, T. DONOVAN, AND J. BLAKE. 1993. Habitat fragmentation in the temperate zone: a perspective for managers. Pp. 331–338 *in* D. M. Finch and P. W. Stangel (editors). Status and management of neotropical migratory birds. USDA Forest Service Gen. Tech. Rep. RM-229. USDA Forest Service Rocky Mountain Forest and Range Experiment Station, Ft. Collins, CO.

FAABORG, J., AND J. TERBORGH. 1980. Patterns of migration in the West Indies. Pp. 157–163 *in* A. Keast and E. S. Morton (editors). Migrant birds in the Neotropics: ecology, behavior, distribution, and conservation. Smithsonian Institution Press, Washington, D.C.

FARLEY, G. H., L. M. ELLIS, J. N. STUART, AND N. J. SCOTT, JR. 1994. Avian species richness in different-aged stands of riparian forest along the middle Rio Grande, New Mexico. Conservation Biology 8: 1098–1108.

FASOLA, M., AND F. FRATICELLI. 1990 Non-competitive habitat use by foraging passerines during spring migrations. Ethology, Ecology, and Evolution 2:363–371.

FERNS, P. N. 1975. Feeding behavior of autumn passage migrants in northeast Portugal. Ringing and Migration 1:3–11.

FICKEN, M. S. 1962. Agonistic behavior and territory in the American Redstart. Auk 79:607–632.

FICKEN, M. S., AND R. F. FICKEN. 1962. The comparative ethology of the wood warblers: a review. Living Bird 1:103–122.

FINCH, D. M. 1989. Species abundances, guild dominance patterns, and community structure of breeding riparian birds. Pp. 629–645 *in* R. R. Sharitz and J. W. Gibbons (editors). Freshwater wetlands and wildlife: perspectives on natural, managed, and degraded ecosystems. DOE-CONF 8603101, Office of Scientific and Technical Information, U.S. Department of Energy, Oakridge, TN.

FINCH, D. M., AND L. F. RUGGIERO. 1993. Wildlife habitats and biological diversity in the Rocky Mountains and northern Great Plains. Natural Areas Journal 13:191–203.

FINCH, D. M., AND P. W. STANGEL (EDITORS). 1993. Status and management of neotropical migratory birds. USDA Forest Service Gen. Tech. Rep. RM-229. USDA Forest Service Rocky Mountain Forest and Range Experiment Station, Ft. Collins, CO.

FINCH, D. M., AND J. A. TAINTER (EDITORS). 1995a. Ecology and sustainability of the Middle Rio Grande Basin. USDA Forest Service Gen. Tech. Rep. RM-268. USDA Forest Service Rocky Mountain Forest and Range Experiment Station, Fort Collins, CO.

FINCH, D. M., AND J. A. TAINTER. 1995b. Introduction: ecosystem research in a human context. Pp. 1–11 *in* D. M. Finch and J. A. Tainter (editors). Ecology and sustainability of the Middle Rio Grande Basin. USDA Forest Service Gen. Tech. Rep. RM-268. USDA Forest Service Rocky Mountain Forest and Range Experiment Station, Fort Collins, CO.

FINCH, D. M., G. L. WOLTERS, W. YONG, AND M. J. MUND. 1995. Plants, arthropods, and birds of the Rio Grande. Pp. 133–164 *in* D. M. Finch and J. A. Tainter (editors). Ecology and sustainability of the Middle Rio Grande Basin. USDA Forest Service Gen. Tech. Rep. RM-268. USDA Forest Service Rocky Mountain Forest and Range Experiment Station, Fort Collins, CO.

FITZPATRICK, J. W. 1980. Wintering of North American tyrant flycatchers in the Neotropics. Pp. 67–78 *in* A. Keast and E. S. Morton (editors). Migrant birds in the Neotropics: ecology, behavior, distribution, and conservation. Smithsonian Institution Press, Washington, D.C.

FLATHER, C. H., AND L. A. JOYCE. 1994. Species endangerment patterns in the United States. USDA Forest Service Gen. Tech. Rep. RM-241. USDA Forest Service Rocky Mountain Forest and Range Experiment Station, Fort Collins, CO.

FORD, H. A., L. HUDDY, AND H. BELL. 1990. Seasonal changes in foraging behavior of three passerines in Australian eucalyptus woodland. Studies in Avian Biology 13:245–253.

FORMAN, R. T. T., A. E. GALLI, AND C. F. LECK. 1976. Forest size and avian diversity in New Jersey woodlots with some land-use implications. Oecologia 26: 1–8.

FOWELLS, H. A. 1965. Silvics of forest trees of the United States. USDA Forest Service Agriculture Handbook No. 271. USDA Forest Service, Washington, D.C.

FRANCIS, C. M., AND F. COOKE. 1986. Differential timing of spring migration in wood warblers (Parulinae). Auk 103:548–556.

FREEHLING, M. D. 1982. Riparian woodlands of the middle Rio Grande valley, New Mexico: a study of bird populations and vegetation with special reference to Russian olive (*Elaeagnus angustifolia*). U.S. Fish and Wildlife Service, Office of Environment, Region 2, Albuquerque, NM, Contract No. 14-16-0002-78-129. Department of Fisheries and Wildlife Sciences, New Mexico State University, Las Cruces, NM.

FRETWELL, S. D. 1980. Evolution of migration in relation to factors regulating bird numbers. Pp. 517–527 *in* A. Keast and E. S. Morton (editors). Migrant birds in the Neotropics: ecology, behavior, distribution, and conservation. Smithsonian Institution Press, Washington, D.C.

FRETWELL, S. D., AND H. L. LUCAS, JR. 1970. On territorial behavior and other factors influencing habitat distribution in birds. I. Theoretical development. Acta Biotheoretica 19:16–36.

FRY, C. H., J. S. ASH, AND I. J. FERGUSON-LEES. 1970. Spring weights of some palearctic migrants at Lake Chad. Ibis 112:58–82.

GATES, J. E., AND L. W. GYSEL. 1978. Avian nest dispersion and fledging success in field-forest ecotones. Ecology 59:871–883.

GAUTHREAUX, S. A., JR. 1971. A radar and direct visual study of passerine spring migration in southern Louisiana. Auk 88:343–365.

GAUTHREAUX, S. A., JR. 1972. Behavioral responses of migrating birds to daylight and darkness: a radar and direct visual study. Wilson Bulletin 84:136–148.

GAUTHREAUX, S. A., JR. 1978. The ecological significance of behavioral dominance. Pp. 17–54 *in* P. P. G. Bateson and P. H. Klopfer (editors). Perspectives in ethology, vol. 3. Plenum Press, New York, NY.

GAUTHREAUX, S. A., JR. 1979. Priorities in bird migration studies. Auk 96:813–815.

GAUTHREAUX, S. A., JR. 1982a. Age-dependent orientation in migratory birds. Pp. 68–74 *in* F. Papi and H. G. Wallraff (editors). Avian navigation. Springer-Verlag, Berlin, FRG.

GAUTHREAUX, S. A., JR. 1982b. The ecology and evolution of avian migration systems. Pp. 493–516 *in* D. S. Farner and J. R. King (editors). Avian biology, vol. VI. Academic Press, New York, NY.

GAUTHREAUX, S. A., JR. 1991. The flight behavior of migrating birds in changing wind fields: radar and visual analyses. American Zoologist 31:187–204.

GOSS-CUSTARD, J. D., AND S. E. A. LE V. DIT DURELL. 1983. Individual and age differences in the feeding ecology of Oystercatchers *Haematopus ostralegus* wintering on the Exe Estuary, Devon. Ibis 125:155–171.

GOSSELINK, J. G., C. L. CORDES, AND J. W. PARSONS. 1979. An ecological characterization study of the Chenier Plain coastal ecosystem of Louisiana and Texas. FWS/OBS-78/9 through 78/11. US Fish and Wildlife Service, Office of Biological Services, Washington, D.C.

GOTMARK, F., AND U. UNGER. 1994. Are conspicuous birds unprofitable prey? Field experiments with hawks and stuffed prey species. Auk 111:251–262.

GRABER, J. W., AND R. R. GRABER. 1983. Feeding rates of warblers in spring. Condor 85:139–150.

GREENBERG, R. S. 1979. Body size, breeding habitat, and winter exploitation systems in *Dendroica.* Auk 96:756–766.

GREENBERG, R. S. 1980. Demographic aspects of long-distance migration. Pp. 493–516 *in* A. Keast and E. S. Morton (editors). Migrant birds in the Neotropics: ecology, behavior, distribution, and conservation. Smithsonian Institution Press, Washington, D.C.

GREENBERG, R. S. 1981. Frugivory in some migrant tropical forest wood warblers. Biotropica 13:215–223.

GREENBERG, R. S. 1983. The role of neophobia in determining foraging specialization of some migratory warblers. American Naturalist 122:444–453.

GREENBERG, R. S. 1984a. The winter exploitation systems of Bay-breasted and Chestnut-sided warblers in Panama. University of California Publications in Zoology 116:1–107.

GREENBERG, R. S. 1984b. Differences in feeding neophobia in tropical migrant warblers, *Dendroica castanea* and *D. pensylvanica.* Journal of Comparative Psychology 98:131–136.

GREENBERG, R. S. 1984c. Neophobia in the foraging-site selection of a neotropical migrant bird: an experimental study. Proceedings of the National Academy of Sciences (USA) 81:3778–3780.

GREENBERG, R. S. 1985. A comparison of foliage discrimination learning in a specialist and a generalist species of migrant wood warbler (Aves: Parulidae). Canadian Journal of Zoology 63:773–776.

GREENBERG, R. S. 1986. Competition in migrant birds during the nonbreeding season. Current Ornithology 3:281–307.

GREENBERG, R. S. 1987a. Seasonal foraging specialization in the Worm-eating Warbler. Condor 89:158–168.

GREENBERG, R. S. 1987b. Social facilitation does not reduce neophobia in Chestnut-sided Warblers (*Dendroica pensylvania*). Journal of Ethology 5:7–10.

GREENBERG, R. S. 1990. Ecological plasticity, neophobia, and resource use in birds. Studies in Avian Biology 13:431–437.

GREENBERG, R. S. 1992. Forest migrants in non-forest habitats on the Yucatan Peninsula. Pp. 273–286 *in* J. M. Hagan, III, and D. W. Johnston (editors). Ecology and conservation of neotropical migrant landbirds. Smithsonian Institution Press, Washington, D.C.

GREENBERG, R. S. 1993. Frugivory and coexistence in a resident and migratory vireo on the Yucatan Peninsula. Condor 95:990–999.

GREENBERG, R. S., M. S. FOSTER, AND L. MARQUEZ-VALDELAMAR. 1995a. The role of the white-eyed vireo in the dispersal of *Bursera* fruit on the Yucatan Peninsula. Journal of Tropical Ecology 11:619–639.

GREENBERG, R. S., AND J. A. GRADWOHL. 1980. Observations of paired Canada Warblers *Wilsonia canadensis* during migration in Panama. Ibis 122:509–512.

GREENBERG, R. S., AND S. LUMPKIN. 1991. Birds over troubled forests. Smithsonian Migratory Bird Center, National Zoological Park, Washington, D.C.

GREENBERG, R. S., J. SALGADO, I. WARKENTIN, AND P. BICHIER. 1995b. Managed forest patches and the conservation of migratory birds in Chiapas, Mexico. Pp. 178–189 *in* M. H. Wilson and S. A. Sader (editors). Conservation of neotropical migratory birds in Mexico. Maine Agricultural and Forest Experiment Station, Catemaco, Mexico.

GRUBB, T. C., JR., AND L. GREENWALD. 1982. Sparrows and a brushpile: foraging response to different combinations of predation risk and energy cost. Animal Behavior 32:120–131.

GRUBB, T. C., JR., AND M. S. WOODREY. 1990. Sex, age, intraspecific dominance status, and the use of food by birds wintering in temperate-deciduous and cold-coniferous woodlands: a review. Studies in Avian Biology 13:270–279.

GWINNER, E. 1986. Circadian rhythms. Springer-Verlag, Berlin, FRG.

GWINNER, E., H. BIEBACH, AND I. V. KRIES. 1985. Food availability affects migratory restlessness in caged Garden Warblers (*Sylvia borin*). Naturwissenshaften 72:51–52.

HAGAN, J. M., III, AND D. W. JOHNSTON (EDITORS). 1992. Ecology and conservation of neotropical migrant landbirds. Smithsonian Institution Press, Washington, D.C.

HALL, G. A., AND R. K. BELL. 1981. The diurnal migration of passerines along an Appalachian ridge. American Birds 35:135–138.

HAMEL, P. B. 1992. Land manager's guide to the birds of the South. The Nature Conservancy, Chapel Hill, NC.

HANSSON, L. 1983. Bird numbers across edges between

mature conifer forest and clearcuts in Central Sweden. Ornis Scandinavica 14:97–103.

HANSSON, M., AND J. PETTERSSON. 1989. Competition and fat deposition in Goldcrests (*Regulus regulus*) at a migration stopover site. Vogelwarte 35:21–31.

HAYES, F. E. 1995. Definitions for migrant birds: what is a neotropical migrant? Auk 112:521–523.

HEBRARD, J. J. 1978. Habitat selection in two species of *Spizella*: a concurrent laboratory and field study. Auk 95:404–410.

HEDENSTRÖM, A., S. BENSCH, D. HASSELQUIST, M. LOCKWOOD, AND U. OTTOSSON. 1993. Migration, stopover and moult of the Great Reed Warbler *Acrocephalus arundinaceus* in Ghana, West Africa. Ibis 135:177–180.

HEHNKE, M., AND C. P. STONE. 1979. Value of riparian vegetation to avian populations along the Sacramento River system. Pp. 228–235 *in* R. Johnson and J. F. McCormick (editors). Strategies for protection and management of floodplain wetlands and other riparian ecosystems. USDA Forest Service. Gen. Tech. Rep. WO-12. USDA Forest Service, Washington, D.C.

HELMS, C. W., AND W. H. DRURY. 1960. Winter and migratory weight and fat: field studies on some North American buntings. Bird-Banding 31:1–40.

HERRERA, C. M. 1981. Fruit variation and competition for dispersers in natural populations of *Smilax aspera*. Oikos 36:51–58.

HERRERA, C. M. 1984. A study of avian frugivores, bird-dispersed plants, and their interaction in mediterranean scrublands. Ecological Monographs 54:1–23.

HERRERA, C. M. 1985. Habitat-consumer interactions in frugivorous birds. Pp. 341–367 *in* M. L. Cody (editor). Habitat selection in birds. Academic Press, San Diego, CA.

HERRERA, C. M. 1987. Vertebrate-dispersed plants of the Iberian Peninsula: a study of fruit characteristics. Ecological Monographs 57:305–331.

HIEBERT, S. 1991. Seasonal differences in the response of rufous hummingbirds to food restriction: body mass and the use of torpor. Condor 93:526–537.

HILDÉN, O. 1965. Habitat selection in birds. Annales Zoologici Fennici 2:53–75.

HILTY, S. L. 1980. Relative abundance of north temperate zone breeding migrants in western Colombia and their impact at fruiting trees. Pp. 265–271 *in* A. Keast and E. S. Morton (editors). Migrant birds in the Neotropics: ecology, behavior, distribution, and conservation. Smithsonian Institution Press, Washington, D.C.

HINE, J. S. 1906. A second contribution to the entomology of the region of the Gulf Biologic Station. Gulf Biologic Station Bulletin 6:65–83.

HINK, V. C., and R. D. Ohmart. 1984. Middle Rio Grande Biological Survey. Final Report. Army Corps of Engineers Contract No. DACW47-81-C-0015. Center for Environmental Studies, Arizona State University, Tempe, AZ.

HODORFF, R. A., C. H. SIEG, AND R. L. LINDER. 1988. Wildlife response to stand structure of deciduous woodlands. Journal of Wildlife Management 82: 667–673.

HOFFMAN, S. W. 1990. Bosque biological monitoring program: bird population surveys in Rio Grande Valley State Park (1987–1990). Final Report. City of Albuquerque, Open Space Division, Albuquerque, NM.

HOLMES, R. T., AND T. W. SHERRY. 1988. Assessing population trends of New Hampshire forest birds: local vs. regional patterns. Auk 105:756–768.

HOLMES, R. T., T. W. SHERRY, AND L. REITSMA. 1989. Population structure, territoriality, and overwinter survival of two migrant warbler species in Jamaica. Condor 91:545–561.

HOWE, H. F. 1977. Bird activity and seed dispersal of a tropical wet forest tree. Ecology 58:539–550.

HOWE, H. F. 1981. Dispersal of a neotropical nutmeg (*Virola sebifora*) by birds. Auk 98:88–98.

HOWE, H. F., AND D. DESTEVEN. 1979. Fruit production, migrant bird visitation, and seed dispersal of *Guarea glabra* in Panama. Oecologia 39:185–196.

HOWE, R. W. 1984. Local dynamics of bird assemblages in small forest habitat islands in Australia and North America. Ecology 65:1585–1601.

HOWE, W. H., AND F. L. KNOPF. 1991. On the imminent decline of Rio Grande cottonwoods in central New Mexico. Southwestern Naturalist 36:218–224.

HUBBARD, J. P. 1971. The summer birds of the Gila Valley, New Mexico. Nemouria 2:1–35.

HUNTER, W. C., M. F. CARTER, D. N. PASHLEY, AND K. BARKER. 1993. The Partners in Flight species prioritization scheme. Pp. 109–119 *in* D. M. Finch and P. W. Stangel (editors). Status and management of neotropical migratory birds. USDA Forest Service Gen. Tech. Rep. RM-229. USDA Forest Service Rocky Mountain Forest and Range Experiment Station, Ft. Collins, CO.

HUNTER W. C., R. D. OHMART, AND B. W. ANDERSON. 1988. Use of exotic saltcedar (*Tamarix chinensis*) by birds in arid riparian systems. Condor 90:113–123.

HUSSELL, D. J. T. 1980. The timing of fall migration and molt in Least Flycatchers. Bird-Banding 51:65–71.

HUSSELL, D. J. T. 1981. Migrations of the Least Flycatcher in southern Ontario. Journal of Field Ornithology 52:97–111.

HUSSELL, D. J. T. 1982. The timing of fall migration in Yellow-bellied Flycatchers. Journal of Field Ornithology 53:1–6.

HUSSELL, D. J. T. 1991. Fall migrations of Alder and Willow flycatchers in southern Ontario. Journal of Field Ornithology 62:260–270.

HUSSELL, D. J. T., T. DAVIS, AND R. D. MONTGOMERIE. 1967. Differential fall migration of adult and immature Least Flycatchers. Bird-Banding 38:61–66.

HUTTO, R. L. 1980. Winter habitat distribution of migratory land birds in western Mexico, with special reference to small foliage-gleaning insectivores. Pp. 181–203 *in* A. Keast and E. S. Morton (editors). Migrant birds in the Neotropics: ecology, behavior, distribution, and conservation. Smithsonian Institution Press, Washington, D.C.

HUTTO, R. L. 1981. Seasonal variation in the foraging behavior of some migratory western wood warblers. Auk 98:765–777.

HUTTO, R. L. 1985a. Seasonal changes in the habitat distribution of transient insectivorous birds in southeastern Arizona: competition mediated? Auk 102: 120–132.

HUTTO, R. L. 1985b. Habitat selection by nonbreeding, migratory land birds. Pp. 455–476 *in* M. L. Cody (editor). Habitat selection in birds. Academic Press, San Diego, CA.

HUTTO, R. L. 1988. Is tropical deforestation responsible for the reported declines in neotropical migrant populations? American Birds 42:375–379.

IZHAKI, I. 1992. A comparative analysis of the nutritional diet of mixed and exclusive fruit diets for Yellow-vented Bulbuls. Condor 94:912–923.

IZHAKI, I., AND U. N. SAFRIEL. 1985. Why do fleshy-fruit plants of the Mediterranean scrub intercept fall—but not spring—passage migratory birds? Oecologia 67:40–43.

IZHAKI, I., AND U. N. SAFRIEL. 1989. Why are there so few exclusively frugivorous birds? Experiments on fruit digestibility. Oikos 54:23–32.

IZHAKI, I., AND U. N. SAFRIEL. 1990. Weight losses due to exclusive fruit diet—interpretation and evolutionary implications: a reply to Mack and Sedinger. Oikos 57:140–142.

JAMES, F. C. 1971. Ordinations of habitat relationships among breeding birds. Wilson Bulletin 83:215–236.

JAMES, F. C., R. F. JOHNSTON, N. O. WARNER, G. J. NIEMI, AND W. J. BOECKLIN. 1984. The Grinnellian niche of the Wood Thrush. American Naturalist 124: 17–47.

JANDER, R. 1975. Ecological aspects of spatial orientation. Annual Review of Ecology and Systematics 6:171–188.

JOHNSON, D. L. 1980. The comparison of usage and availability measurements for evaluating resource preference. Ecology 61:65–71.

JOHNSON, N. K. 1965. Differential timing and routes of spring migration in the Hammond's Flycatcher. Condor 67:423–437.

JOHNSON, N. K. 1973. Spring migration of the Western Flycatcher with notes of seasonal changes in sex and age ratios. Bird-Banding 44:205–220.

JOHNSON, N. K. 1974. Molt and age determination in Western and Yellowish Flycatchers. Auk 91:111–131.

JOHNSON, R. A., M. F. WILLSON, J. M. THOMPSON, AND R. I. BERTIN. 1985. Nutritional values of wild fruits and consumption by migrant frugivorous birds. Ecology 66:819–827.

JOHNSON, T. W. 1993. Involving the private sector in Georgia's conservation initiatives for neotropical birds. Pp. 45–46 *in* D. M. Finch and P. W. Stangel (editors). Status and management of neotropical migratory birds. USDA Forest Service Gen. Tech. Rep. RM-229. USDA Forest Service Rocky Mountain Forest and Range Experiment Station, Ft. Collins, CO.

JONES, G. A., AND C. J. NORMENT. 1998. Absence of breeding readiness in neotropical and long-distance migrant landbirds during spring stopover. Condor 100:373–376.

JORDANO, P. 1981. Alimentacion y relaciones troficas entre los passeriformes en paso otonal por una localidad de Andalucia central. Doñana Acta Vertebrata 8:103–124.

JORDANO, P. 1987. Frugivory, external morphology and digestive system in Mediterranean sylviid warblers *Sylvia* spp. Ibis 129:175–189.

JORDANO, P. 1988. Diet, fruit-choice, and variation in body condition of frugivorous warblers in Mediterranean scrubland. Ardea 76:193–209.

JORDANO, P., AND C. M. HERRERA. 1981. The frugivorous diet of Blackcap populations *Sylvia atricapilla* wintering in southern Spain. Ibis 123:502–507.

KARLSSON, L., K. PERSSON, J. PETTERSSON, AND G. WALINDER. 1988. Fat-weight relationships and migratory strategies in the Robin *Erithacus rubecula* at two stop-over sites in south Sweden. Ringing and Migration 9:160–168.

KEAST, A., AND E. S. MORTON (EDITORS). 1980. Migrant birds in the Neotropics: ecology, behavior, distribution, and conservation. Smithsonian Institution Press, Washington, D.C.

KELSEY, M. G. 1992. Conservation of migrants on their wintering grounds: an overview. Ibis 134 (suppl. 1): 100–112.

KERLINGER, P. 1989. Flight strategies of migrating hawks. University of Chicago Press, Chicago, IL.

KERLINGER, P., AND F. R. MOORE. 1989. Atmospheric structure and avian migration. Current Ornithology 6:109–142.

KERLINGER, P., AND D. WIEDNER. 1991. The economics of birding at Cape May, New Jersey. Pp. 324–334 in J. A. Kusler (editor). Ecotourism and resource conservation, 2nd international symposium: ecotourism and resource conservation. Ecotourism and Resource Conservation Project, Miami Beach, FL.

KETTERSON, E. D., AND V. NOLAN, JR. 1982. The role of migration and winter mortality in the life history of a temperate-zone migrant, the Dark-eyed Junco, as determined from demographic analyses of winter populations. Auk 99:243–259.

KETTERSON, E. D., AND V. NOLAN, JR. 1983. The evolution of differential bird migration. Current Ornithology 3:357–402.

KETTERSON, E. D., AND V. NOLAN, JR. 1985. Intraspecific variation in avian migration patterns. Pp. 553–579 *in* M. A. Rankin (editor). Migration: mechanisms and adaptive significance. Contributions in Marine Science, vol. 27. University of Texas, Austin, TX.

KETTERSON, E. D., AND V. NOLAN, JR. 1988. A possible role for experience in the regulation of the timing of bird migration. Acta XIX Congressus Internationalis 19:2169–2179.

KING, J. R. 1974. Seasonal allocation of time and energy resources in birds. Pp. 4–85 in R. A. Paynter, Jr. (editor). Avian energetics. Publications of the Nuttall Ornithological Club No. 15.

KLOPFER, P. H. 1967. Behavioral stereotypy in birds. Wilson Bulletin 79:290–300.

KLOPFER, P., AND J. HAILMAN. 1965. Habitat selection in birds. Advances in the Study of Behavior 1:279–303.

KLOPFER, P. H., AND R. H. MACARTHUR. 1960. Niche size and faunal diversity. American Naturalist 94: 293–300.

KNOPF, F. L., R. R. JOHNSON, T. RICH, F. B. SAMSON, AND R. C. SZARO. 1988. Conservation of riparian ecosystems in the United States. Wilson Bulletin 12: 289–298.

KNOPF, F. L., J. A. SEDGWICK, AND R. W. CANNON. 1988. Guild structure of a riparian avifauna relative to seasonal cattle grazing. Journal of Wildlife Management 52:280–290.

KOIVULA, K., K. LAHTI, S. RYTKÖNEN, AND M. ORELL. 1994. Do subordinates expose themselves to predation? Field experiments on feeding site selection by Willow Tits. Journal of Avian Biology 25:178–183.

KUENZI, A. J., F. R. MOORE, AND T. R. SIMONS. 1991. Stopover of neotropical landbird migrants following trans-Gulf migration. Condor 93:869–883.

LACK, D. 1933. Habitat selection in birds with special reference to the effects of afforestation on the Breckland avifauna. Journal of Animal Ecology 2:239–262.

LACK, D. 1946. Do juvenile birds survive less well than adults? British Birds 32:258–264.

LACK, D., AND P. LACK. 1973. Wintering warblers in Jamaica. Living Bird 11:129–153.

LAURSEN, K. 1978. Interspecific relationships between some insectivorous passerine species, illustrated by their diet during spring migration. Ornis Scandinavica 9:178–192.

LAVEE, D., AND U. N. SAFRIEL. 1989. The dilemma of cross-desert migrants—stopover or skip a small oasis? Journal of Arid Environments 17:69–81.

LAVEE, D., U. N. SAFRIEL, AND I. MEILIJSON. 1991. For how long do trans-Saharan migrants stop over at an oasis? Ornis Scandinavica 22:33–44.

LEAL, D. A., R. A. MEYER, AND B. C. THOMPSON. 1996. Avian community composition and habitat importance in the Rio Grande corridor of New Mexico. Pp. 62–68 *in* D. W. Shaw and D. M. Finch (technical coordinators). Desired future conditions for southwestern riparian ecosystems: bringing interests and concerns together. USDA Forest Service Gen. Tech. Rep. RM-272. USDA Forest Service Rocky Mountain Forest and Range Experiment Station, Fort Collins, CO.

LEBERG, P. L., T. J. SPENGLER, AND W. C. BARROW, JR. 1996. Lipid and water depletion in migrating passerines following passage over the Gulf of Mexico. Oecologia 106:1–7.

LEBERMAN, R. C., AND M. H. CLENCH. 1973. Birdbanding at Powdermill, 1972. Research Rep. No. 31, Carnegie Museum of Natural History, Pittsburgh, PA.

LEBERMAN, R. C., AND M. H. CLENCH. 1975. Bird-Banding at Powdermill, 1974. Research Rep. No. 36, Carnegie Museum of Natural History, Pittsburgh, PA.

LECK, C. F. 1972. The impact of some North American migrants at fruiting trees in Panama. Auk 89:842–850.

LEISLER, B., AND H. WINKLER. 1985. Ecomorphology. Current Ornithology 2:155–186.

LEVEY, D. J. 1988. Tropical wet forest treefall gaps and distributions of understory birds and plants. Ecology 69:1076–1089.

LEVEY, D. J. 1990. Habitat-dependent fruiting behaviour of an understorey tree, *Miconia centrodesma,* and tropical treefall gaps as keystone habitats in Costa Rica. Journal of Tropical Ecology 6:409–420.

LEVEY, D. J. 1991. Digestive processing of fruits and its consequences for fruit-frugivore coevolution. Acta XX Congressus Internationalis Ornithologici 20:1624–1629.

LEVEY, D. J. 1994. Why we should adopt a broader view of neotropical migrants. Auk 111: 233–236.

LEVEY, D. J., AND G. E. DUKE. 1992. How do frugivores process fruit? Gastrointestinal transit and glucose absorption in cedar waxwings (*Bombycilla cedrorum*). Auk 109:722–730.

LEVEY, D. J., AND W. H. KARASOV. 1989. Digestive responses of temperate birds switched to fruit or insect diets. Auk 106:675–686.

LEVEY, D. J., AND W. H. KARASOV. 1994. Gut passage of insects by European Starlings and comparisons with other species. Auk 111:478–481.

LEVEY, D. J., AND F. G. STILES. 1992. Evolutionary precursors of long-distance migration: resource availability and movement patterns in neotropical landbirds. American Naturalist 140:447–476.

LEVINS, R. 1968. Evolution in changing environments. Princeton University Press, Princeton, NJ.

LIMA, S. L., K. L. WIEBE, AND L. M. DILL. 1987. Protective cover and the use of space by finches: is closer better? Oikos 50:225–230.

LINCOLN, F. C. 1935. The migration of North American birds. U.S. Department of Agriculture, Circular No. 363, Washington, D.C.

LINDSTRÖM, Å. 1989. Finch flock size and risk of hawk predation at a migratory stopover site. Auk 106: 225–232.

LINDSTRÖM, Å. 1990a. Stopover ecology of migrating birds. Ph.D. dissertation. University of Lund, Lund, Sweden.

LINDSTRÖM, Å. 1990b. The role of predation risk in stopover habitat selection in migrating Bramblings *Fringilla montifringilla.* Behavioral Ecology 1:102–106.

LINDSTRÖM, Å. 1991. Maximum fat deposition rates in migrating birds. Ornis Scandinavica 22:12–19.

LINDSTRÖM, Å., AND T. ALERSTAM. 1986. The adaptive significance of reoriented migration of Chaffinches *Fringilla coelebs* and Bramblings *F. montifringilla* during autumn in southern Sweden. Behavioral Ecology and Sociobiology 19:417–424.

LINDSTRÖM, Å., D. HASSELQUIST, S. BENSCH, AND M. GRAHN. 1990. Asymmetric contests over resources for survival and migration: a field experiment with Bluethroats. Animal Behavior 40:453–461.

LINDSTRÖM, Å., AND T. PIERSMA. 1993. Mass changes in migratory birds: the evidence for fat and protein storage reexamined. Ibis 135:70–78.

LIVINGSTON, M. F., AND S. D. SCHEMNITZ. 1996. Summer bird/vegetation associations in tamarisk and native habitat along the Pecos River, southeastern New Mexico. Pp. 171–180 *in* D. W. Shaw and D. M. Finch (technical coordinators). Desired future conditions for southwestern riparian ecosystems: bringing interests and concerns together. USDA Forest Service Gen. Tech. Rep. RM-272. USDA Forest Service Rocky Mountain Forest and Range Experiment Station, Fort Collins, CO.

LORIA, D. E., AND F. R. MOORE. 1990. Energy demands of migration on Red-eyed Vireos *Vireo olivaceus.* Behavioral Ecology 1:24–35.

LOVEI, G. L. 1989. Passerine migration between the

Palaearctic and Africa. Current Ornithology 6:143–174.

LOVEJOY, T. E. 1983. Tropical deforestation and North American migrant birds. Biological Conservation 1:126–128.

LOVETTE, I. J., AND R. T. HOLMES. 1995. Foraging behavior of American Redstarts in breeding and wintering habitats: implications for relative food availability. Condor 97:782–791.

LOWERY, G. H., JR. 1945. Trans-Gulf migration of birds and the coastal hiatus. Wilson Bulletin 57:92–121.

LOWERY, G. H., JR. 1951. A quantitative study of the nocturnal migration of birds. University of Kansas Publications, Museum of Natural History 3:361–472.

LOWERY, G. H., JR. 1974. Louisiana Birds. Louisiana State University Press, Baton Rouge, LA.

LYNCH, J. F., E. S. MORTON, AND M. E. VAN DER VOORT. 1985. Habitat segregation between the sexes of wintering Hooded Warblers (*Wilsonia citrina*). Auk 102:714–721.

LYNCH, J. F., AND D. F. WHIGHAM. 1984. Effects of forest fragmentation on breeding bird communities in Maryland, USA. Biological Conservation 28:287–324.

MABEY, S. E., J. MCCANN, L. J. NILES, C. BARTLETT, AND P. KERLINGER. 1993. The migratory songbird coastal corridor final report. Report No. NA90AA-H-CZ839, Virginia Department of Environmental Quality Coastal Resources Management Program, Richmond, VA.

MACARTHUR, R. H. 1972. Geographical ecology. Harper and Row, New York, NY.

MACARTHUR, R. H., AND J. W. MACARTHUR. 1961. On bird species diversity. Ecology 42:594–598.

MACARTHUR, R. H., AND E. O. WILSON. 1967. The theory of island biogeography. Princeton University Press, Princeton, NJ.

MACCLINTOCK, L., R. F. WHITCOMB, AND B. L. WHITCOMB. 1977. Island biogeography and "habitat islands" of eastern forest. II. Evidence of the value of corridors and minimization of isolation in preservation of biotic diversity. American Birds 31:6–16.

MACK, A. L. 1990. Is frugivory limited by secondary compounds in fruits? Oikos 57:135–138.

MAGEE, M. J. 1924. Notes on the Purple Finch. Auk 41:606–610.

MALMBORG, P. K., AND M. F. WILLSON. 1988. Foraging ecology of avian frugivores and some consequences for seed dispersal in an Illinois woodlot. Condor 90:173–186.

MANGEL, M., AND C. W. CLARK. 1988. Dynamic modeling in behavioral ecology. Princeton University Press, Princeton, NJ.

MANGEL, M., L. M. TALBOT, G. K. MEFFE, M. T. AGARDY, D. L. ALVERSON, J. BARLOW, D. B. BOTKIN, G. BUDOWSKI, T. CLARK, J. COOKE, R. H. CROZIER, P. K. DAYTON, D. L. ELDER, C. W. FOWLER, S. FUNTOWICZ, J. GISKE, R. J. HOFMAN, S. J. HOT, S. R. KELLERT, L. A. KIMBALL, D. LUDWIG, K. MAGNUSSON, B. S. MALAYANG, III, C. MANN, E. A. NORSE, S. P. NORTHRIDGE, W. F. PERRIN, C. PERRINGS, R. M. PETERMAN, G. B. RABB, H. A. REGIER, J. E. REYNOLDS, III, K. SHERMAN, M. P. SISSENWINE, T. D. SMITH, A. STARFIELD, R. J. TAYLOR, M. F. TILLMAN, C. TOFT, J. R. TWISS, JR., J. WILEN, AND T. P. YOUNG. 1996. Principles for the conservation of wild living resources. Ecological Applications 6:338–362.

MANN, C. C., AND M. L. PLUMMER. 1995. California vs. Gnatcatcher. Audubon 97:38–104.

MARCHETTI, K., AND T. PRICE. 1989. Differences in the foraging of juvenile and adult birds: the importance of developmental constraints. Biological Review 64:51–71.

MARRA, P. P., K. A. HOBSON, AND R. T. HOLMES. 1998. Linking winter and summer events in a migratory bird by using stable-carbon isotopes. Science 282:1884–1886.

MARRA, P. P., T. W SHERRY, AND R. T. HOLMES. 1993. Territorial exclusion by a long-distance migrant warbler in Jamaica: a removal experiment with American Redstarts (*Setophaga ruticilla*). Auk 110:565–572.

MARSH, R. L. 1984. Adaptations of the Gray Catbird *Dumetella carolinensis* to long-distance migration: flight muscle hypertrophy associated with elevated body mass. Physiological Zoology 57:105–117.

MARTIN, A. C., H. S. ZIM, AND A. L. NELSON. 1951. American wildlife and plants. Dover Publications, New York, NY.

MARTIN, G. R. 1990. The visual problems of nocturnal migration. Pp. 187–197 *in* E. Gwinner (editor). Bird migration: the physiology and ecophysiology. Springer-Verlag, Berlin, FRG.

MARTIN, T. E. 1980. Diversity and abundance of spring migratory birds using habitat islands on the Great Plains. Condor 82:430–439.

MARTIN, T. E. 1981. Limitation in small habitat islands: chance or competition? Auk 98:715–734.

MARTIN, T. E. 1985. Selection of second-growth woodlands by frugivorous migrating birds in Panama: an effect of fruit size and plant density. Journal of Tropical Ecology 1:157–170.

MARTIN, T. E. 1986. Competition in breeding birds: on the importance of considering processes at the level of the individual. Current Ornithology 4:181–210.

MARTIN, T. E., AND D. M. FINCH (EDITORS). 1995. Ecology and management of neotropical migratory birds: a synthesis and review of critical issues. Oxford University Press, New York, NY.

MARTIN, T. E., AND J. R. KARR. 1986. Patch utilization by migrating birds: resource oriented? Ornis Scandinavica 17:165–174.

MARTIN, T. E., AND J. R. KARR. 1990. Behavioral plasticity of foraging maneuvers of migratory warblers: multiple selection periods for niches? Studies in Avian Biology 13:353–359.

MARTIN, T. E., AND P. A. VOHS. 1978. Configuration of shelterbelts for optimum utilization by birds. Pp. 79–88 *in* R.W. Tinus (editor). Proceedings of the 30th annual Forestry Commission and Great Plains Agricultural Council. Great Plains Agricultural Council Publication No. 87, Tulsa, OK.

MARTINEZ DEL RIO, C., AND W. H. KARASOV. 1990. Digestion strategies in nectar-and fruit-eating birds and the sugar composition of plant rewards. American Naturalist 136:618–637.

MASON, J. R., AND R. F. REIDINGER. 1981. Effects of social facilitation and observational learning on feeding behavior of the Red-winged Blackbird (*Agelaius phoeniceus*). Auk 98:778–784.

Mason, W. 1979. Habitat selection by the Parulidae during spring migration along the South Fork Creek in Glasgow, KY. Kentucky Warbler 55:39–42.

May, R.M. 1994. The effects of spatial scale on ecological questions and answers. Pp. 1–17 *in* P. J. Edwards, R. M. May, and N. R. Webb (editors). Large-scale ecology and conservation biology. Blackwell Scientific Publications, Oxford, UK.

McCann, J. M., S. E. Mabey, L. J. Niles, C. Bartlett, and P. Kerlinger. 1993. A regional study of coastal migratory stopover habitat for neotropical migrant songbirds: land management implications. Transactions of the North American Wildlife and Natural Resources Conference 58:398–407.

McLaren, I. A. 1981. The incidence of vagrant landbirds on Nova Scotian islands. Auk 98:243–257.

Mead, C. 1966. Pre-migratory weights of trans-Saharan migrants. Ringers' Bulletin 2:15–16.

Mehlum, F. 1983. Weight changes in migrating Robins (*Erithacus rubecula*) during stop-over at the island of Store Faerder, Outer Oslofjord, Norway. Fauna Norvegica Series C, Cinclus 6:57–61.

Metcalfe, N. B., and R. W. Furness. 1984. Changing priorities: the effect of pre-migratory fattening on the trade-off between foraging and vigilance. Behavioral Ecology and Sociobiology 15:203–206.

Meyer, W. B., and B. L. Turner, II. 1992. Human population growth and global land-use/cover change. Annual Review of Ecology and Systematics 23:39–61.

Miles, D. B., and R. E. Ricklefs. 1984. The correlation between ecology and morphology in deciduous forest passerines. Ecology 65:1629–1640.

Mitsch, W. J., and J. G. Gosselink. 1986. Wetlands. Van Nostrand Reinhold, New York, NY.

Moermond, T. C., and J. S. Denslow. 1985. Neotropical avian frugivores: patterns of behavior, morphology, and nutrition with consequences for fruit selection. Ornithological Monographs 36:865–897.

Møller, A. P. 1988. Female choice selects for male sexual tail ornaments of the monogamous Swallow. Nature 332:640–642.

Moore, F. R. 1984. Age-dependent variability in the orientation of migratory Savannah Sparrows (*Passerculus sandwichensis*). Auk 101:875–880.

Moore, F. R. 1990a. Evidence for redetermination of migratory direction following wind displacement. Auk 107:425–428.

Moore, F. R. 1990b. Prothonotary Warblers cross the Gulf of Mexico together. Journal of Field Ornithology 61:285–287.

Moore, F. R. 1991a. Ecophysiological and behavioral response to energy demand during migration. Acta XX Congressus Internationalis Ornithologici 20: 753–760.

Moore, F. R. 1991b. Closing remarks: new aspects of avian migration systems. Acta XX Congressus Internationalis Ornithologici 20:787–788.

Moore, F. R. 1994. Resumption of feeding under risk of predation: effect of migratory condition. Animal Behavior 48:975–977.

Moore, F. R., and D. A. Aborn. 1996. Time of departure by Summer Tanagers (*Piranga rubra*) from a stopover site following spring trans-Gulf migration. Auk 113:949–952.

Moore, F. R., S. A. Gauthreaux, Jr., P. Kerlinger, and T. R. Simons. 1993. Stopover habitat: management implications and guidelines. Pp. 58–69 *in* D. M. Finch and P. W. Stangel (editors). Status and management of neotropical migratory birds. USDA Forest Service Gen. Tech. Rep. RM-229. USDA Forest Service Rocky Mountain Forest and Range Experiment Station, Ft. Collins, CO.

Moore, F. R., S. A. Gauthreaux, Jr., P. Kerlinger, and T. R. Simons. 1995. Habitat requirements during migration: important link in conservation. Pp. 121–144 *in* T. E. Martin and D. M. Finch (editors). Ecology and management of neotropical migratory birds: a synthesis and review of critical issues. Oxford University Press, New York, NY.

Moore, F. R., and P. Kerlinger. 1987. Stopover and fat deposition by North American wood-warblers (Parulinae) following spring migration over the Gulf of Mexico. Oecologia 74:47–54.

Moore, F. R., and P. Kerlinger. 1991. Nocturnality, long-distance migration, and ecological barriers. Acta XX Congressus Internationalis Ornithologici 20:1122–1129.

Moore, F. R., P. Kerlinger, and T. R. Simons. 1990. Stopover on a Gulf coast barrier island by spring trans-Gulf migrants. Wilson Bulletin 102:487–500.

Moore, F. R., and M. V. McDonald. 1993. On the possibility that intercontinental landbird migrants copulate *en route*. Auk 110:157–160.

Moore, F. R., and P. Simm. 1985. Migratory disposition and choice of diet by the Yellow-rumped Warbler (*Dendroica coronata*). Auk 102:820–826.

Moore, F. R., and T. R. Simons. 1992a. Habitat suitability and stopover ecology of neotropical landbird migrants. Pp. 345–355 *in* J. M. Hagan, III, and D. W. Johnston (editors). Ecology and conservation of neotropical migrant landbirds. Smithsonian Institution Press, Washington, D.C.

Moore, F. R., and T. R. Simons. 1992b. Use of stopover habitats along the northern coast of the Gulf of Mexico by neotropical landbird migrants. Annual report to the U.S. Fish and Wildlife Service, National Wetlands Research Center, Slidell, LA.

Moore, F. R., and M. S. Woodrey. 1993. Stopover habitat and its importance in the conservation of landbird migrants. Proceedings of the Annual Conference of the Southeastern Association of Fish and Wildlife Agencies 47:447–459.

Moore, F. R., and W. Yong. 1991. Evidence of food-based competition among passerine migrants during stopover. Behavior Ecology and Sociobiology 28: 85–90.

Moreau, R. E. 1969. Comparative weights of some trans-Saharan migrants at intermediate points. Ibis 111:621–624.

Moreau, R. E. 1972. The Palearctic-African bird migration systems. Academic Press, New York, NY.

Morris, S. R., D. W. Holmes, and M. E. Richmond. 1996. A ten-year study of the stopover patterns of migratory passerines during fall migration on Appledore Island, Maine. Condor 98:395–409.

Morris, S. R., M. E. Richmond, and D. W. Holmes. 1994. Patterns of stopover by warblers during spring and fall migration on Appledore Island, Maine. Wilson Bulletin 106:703–718.

MORRISON, M. L., B. G. MARCOT, AND R. W. MANNAN. 1992. Wildlife-habitat relationships: concepts and applications. University of Wisconsin Press, Madison, WI.

MORRISON, M. L., T. A. SCOTT, AND T. TENNANT. 1994. Wildlife habitat restoration in an urban park in southern California. Restoration Ecology 2:17–30.

MORSE, D. H. 1971. The insectivorous bird as an adaptive strategy. Annual Review of Ecology and Systematics 2:177–200.

MORSE, D. H. 1973. Interaction between tit flocks and sparrowhawks *Accipiter nisus*. Ibis 115:591–593.

MORSE, D. H. 1976. Variables affecting the density and territory size of breeding spruce-woods warblers. Ecology 57:290–301.

MORSE, D. H. 1980a. Behavioral mechanisms in ecology. Harvard University Press, Cambridge, MA.

MORSE, D. H. 1980b. Population limitations: breeding or wintering grounds? Pp. 437–453 *in* A. Keast and E. S. Morton (editors). Migrant birds in the Neotropics: ecology, behavior, distribution, and conservation. Smithsonian Institution Press, Washington, D.C.

MORSE, D. H. 1985. Habitat selection in North American Parulid warblers. Pp. 131–157 *in* M. L. Cody (editor). Habitat selection in birds. Academic Press, San Diego, CA.

MORSE, D. H. 1989. American Warblers: an ecological and behavioral perspective. Harvard University Press, Cambridge, MA.

MORTON, E. S. 1973. On the evolutionary advantages and disadvantages of fruit-eating in tropical birds. American Naturalist 107:8–22.

MORTON, E. S. 1980. Adaptations to seasonal changes by migrant land birds in the Panama Canal Zone. Pp. 437–453 *in* A. Keast and E. S. Morton (editors). Migrant birds in the Neotropics: ecology, behavior, distribution, and conservation. Smithsonian Institution Press, Washington, D.C.

MORTON, E. S. 1990. Habitat segregation by sex in the Hooded Warbler: experiments on proximate causation and discussion of its evolution. American Naturalist 135:319–33.

MOSCONI, S. L., AND R. L. HUTTO. 1983. The effect of grazing on the land birds of a western Montana riparian habitat. Pp. 221–233 *in* L. Nelson, J. M. Peek, and P. D. Dalke (editors). Proceedings of the wildlife-livestock relationships symposium. US Forest, Wildlife, and Range Experiment Station, University of Idaho, Moscow, ID.

MOSER, E. B., W. C. BARROW, JR., AND R. B. HAMILTON. 1990. An exploratory use of correspondence analysis to study relationships between avian foraging behavior and habitat. Studies in Avian Biology 13:309–317.

MOUNT, J. M., D. M. FINCH, AND W. J. KRAUSMAN. 1996. Riparian habitat change along the Isleta-Belen reach of the Rio Grande. Pp. 58–61 *in* D. W. Shaw and D. M. Finch (technical coordinators). Desired future conditions for southwestern riparian ecosystems: bringing interests and concerns together. USDA Forest Service Gen. Tech. Rep. RM-272. USDA Forest Service Rocky Mountain Forest and Range Experiment Station, Fort Collins, CO.

MUELLER, A. J., AND N. E. SEARS. 1987. Habitat selection by spring migrants in a Texas coastal woodlot. Bulletin of the Texas Ornithological Society 20:21–26.

MURPHY, D. D. 1989. Conservation and confusion: wrong species, wrong scale, wrong conclusions. Conservation Biology 3:82–84.

MURRAY, B. G., JR. 1966. Migration of age and sex classes of passerines on the Atlantic coast in autumn. Auk 83:352–360.

MURRAY, B. G., JR. 1979. Fall migration of Blackpoll and Yellow-rumped warblers at Island Beach, New Jersey. Bird-Banding 50:1–11.

MURTON, R. K. 1971. The significance of a specific search image in the feeding behaviour of the woodpigeon. Behaviour 40:10–42.

MYERS, J. P., R. I. G. MORRISON, P. E. ANTAS, B. A. HARRINGTON, T. E. LOVEJOY, H. SALLABERRY, S. E. SENNER, AND A. TARAK. 1987. Conservation strategy for migratory species. American Scientist 75:19–26.

NEU, C. W., C. R. BYERS, AND J. M. PEEK. 1974. A technique for analysis of utilization-availability data. Journal of Wildlife Management 38:541–545.

NEWCOMB, W. W., JR. 1961. The Indians of Texas, from prehistoric to modern times. University of Texas Press, Austin, TX.

NISBET, I. C. T., AND L. MEDWAY. 1972. Dispersion, population ecology, and migration of eastern Great Reed Warblers *Acrocephalus orientalis* wintering in Malaysia. Ibis 114:451–494.

NORTHAMPTON COUNTY BOARD OF SUPERVISORS AND THE SUSTAINABLE DEVELOPMENT TASK FORCE. 1994. The sustainable development action strategy for Northampton County, Virginia. Report No. NA270Z0312-01, Virginia Department of Environmental Quality Coastal Resources Management Program, Richmond, VA.

NORTHAMPTON COUNTY DEPARTMENT OF PLANNING AND ZONING. 1989. Information and analysis: comprehensive plan background. Northampton County Department of Planning and Zoning, Eastville, VA.

NORTHAMPTON COUNTY JOINT PLANNING COMMISSION. 1990. Northampton County comprehensive plan. Northampton County Department of Planning and Zoning, Eastville, VA.

NOSS, R. F. 1987. Corridors in real landscapes: a reply to Simberloff and Cox. Conservation Biology 1: 159–164.

NOSS, R. F. 1991. Effects of edge and patchiness on avian habitat use in an old-growth Florida hammock. Natural Areas Journal 11:34–47.

NOSS, R. F., E. T. LAROE, III, AND J. M. SCOTT. 1995. Endangered ecosystems of the United States: a preliminary assessment of loss and degradation. National Biological Service, Biological Report 28.

ODUM, E. P. 1993. Body masses and composition of migrant birds in the eastern United States. Pp. 313–354 *in* J. B. Dunning (editor). CRC handbook of avian body masses. CRC Press, Boca Raton, FL.

ODUM, E. P., C. C. CONNELL, AND H. L. STODDARD. 1961. Flight energy and estimated flight ranges of some migratory birds. Auk 78:515–527.

ODUM, E. P., D. T. ROGERS, AND D. L. HICKS. 1964.

Homeostasis of nonfat components of migratory birds. Science 143:1037–1039.

OHMART, R. D., AND B. W. ANDERSON. 1982. North American desert riparian ecosystems. Pp. 433–479 *in* G. L. Bender (editor). Reference handbook on the deserts of North America. Greenwood Press, Westport, CT.

OLSON, S. L., AND K. E. BLUM. 1968. Avian dispersal of plants in Panama. Ecology 49:565–566.

ORIANS, G. H., AND J. F. WITTENBERGER. 1991. Spatial and temporal scales in habitat selection. American Naturalist 137:S29–S49.

ORMEROD, S. J. 1990. Time of passage, habitat use and mass change of *Acrocephalus* warblers in a South Wales reedswamp. Ringing and Migration 11:1–11.

ORNAT, A. L., AND R. S. GREENBERG. 1990. Sexual segregation by habitat in migratory warblers in Quintana Roo, Mexico. Auk 107:539–543.

OTAHAL, C. D. 1995. Sexual differences in Wilson's Warbler migration. Journal of Field Ornithology 66: 60–69.

OWEN, M., AND J. M. BLACK. 1991. The importance of migration mortality in non-passerine birds. Pp. 360–372 *in* C. M. Perrins, J. D. Lebreton and G. J. M. Hirons (editors). Bird population studies: relevance to conservation and management. Oxford University Press, Oxford, UK.

PALMISANO, A. W. 1967. Ecology of *S. olneyi* and *S. robustus* in Louisiana coastal marshes. M.S. thesis. Louisiana State University, Baton Rouge, LA.

PALMISANO, A. W. 1970. Plant community-soil relationships in Louisiana coastal marshes. Ph.D. dissertation. Louisiana State University, Baton Rouge, LA.

PAMBOUR, B. 1990. Vertical and horizontal distribution of five wetland passerine birds during the postbreeding migration period in a reed-bed of the Camargue, France. Ringing and Migration 11:52–56.

PARKER, T. A., III. 1994. Habitat, behavior, and spring migration of Cerulean Warbler in Belize. American Birds 48:70–75.

PARNELL, J. F. 1969. Habitat relations of the Parulidae during spring migration. Auk 86:505–521.

PARRISH, J. D. 1995a. Effects of needle architecture on warbler habitat selection in a coastal spruce forest. Ecology 76:1813–1820.

PARRISH, J. D. 1995b. Experimental evidence for intrinsic microhabitat preferences in the Black-throated Green Warbler. Condor 97:935–943.

PARRISH, J. D. 1996. Dietary plasticity during migration: evolution and behavioral implications of diet shifts to frugivory in nearctic-neotropical landbird migrants. Ph.D. dissertation. Brown University, Providence, RI.

PARRISH, J. D. 1997. Patterns of frugivory and energetic condition in nearctic landbirds during autumn migration. Condor 99:681–697.

PARRISH, J. D., S. B. COMINGS, M. L. WHITMAN. 1994. A facilitated method for collection of fecal samples from mist-netted birds. North American Bird-Bander 19:49–51.

PARRISH, J. D., AND T. W. SHERRY. 1994. Sexual habitat segregation by American Redstarts wintering in Jamaica: importance of resource seasonality. Auk 111: 38–49.

PARTNERS IN FLIGHT. 1992. Preliminary lists of migrants for Partners in Flight neotropical migratory bird conservation program. Partners in Flight Newsletter 2(1):30.

PARTRIDGE, L. 1979. Habitat selection in titmice. Nature 247:573–574.

PASHLEY, D. N., AND W. C. BARROW, JR. 1992. Effects of land use practices on neotropical migratory birds in bottomland hardwood forests. Pp. 315–320 *in* J. M. Hagan, III, and D. W. Johnston (editors). Ecology and conservation of neotropical migrant landbirds. Smithsonian Institution Press, Washington, D.C.

PEARSON, D. J. 1990. Palaearctic passerine migrants in Kenya and Uganda: temporal and spatial patterns of their movements. Pp. 44–59 *in* E. Gwinner (editor). Bird migration: the physiology and ecophysiology. Springer-Verlag, Berlin, FRG.

PEARSON, S. M., M. G. TURNER, R. H. GARDNER, AND R. V. O'NEILL. 1996. An organism-based perspective of habitat fragmentation. Pp. 77–95 *in* R. C. Szaro and D. W. Johnson (editors). Biodiversity in managed landscapes: theory and practice. Oxford University Press, New York, NY.

PENNINGS, S. C., M. T. NADEAU, AND V. J. PAUL. 1993. Selectivity and growth of the generalist herbivore *Dolabella auricularia* feeding upon complementary resources. Ecology 74:879–890.

PENNYCUICK, C. J. 1975. Mechanics of flight. Pp. 1–75 *in* D. S. Farner and J. R. King (editors). Avian biology, vol. V. Academic Press, New York, NY.

PENNYCUICK, C. J. 1989. Bird flight performance: a practical calculation manual. Oxford University Press, New York, NY.

PENNYCUICK, C. J. 1992. Bird flight performance: a practical calculation manual. Program 1, version 1.1. Oxford University Press, Oxford, UK.

PETIT, D. R. 1991. Habitat associations of migratory birds wintering in Belize, Central America: implications for theory and conservation. Ph.D. dissertation. University of Arkansas, Fayetteville, AR.

PETIT, D. R., J. F. LYNCH, R. L. HUTTO, J. G. BLAKE, AND R. B. WAIDE. 1993. Management and conservation of migratory landbirds overwintering in the Neotropics. Pp. 70–92 *in* D. M. Finch and P. W. Stangel (editors). Status and management of neotropical migratory birds. USDA Forest Service Gen. Tech. Rep. RM-229. USDA Forest Service Rocky Mountain Forest and Range Experiment Station, Ft. Collins, CO.

PETIT, D. R., J. F. LYNCH, R. L. HUTTO, J. G. BLAKE, AND R. B. WAIDE. 1995. Habitat use and conservation in the Neotropics. Pp. 145–197 *in* T. E. Martin and D. M. Finch (editors). Ecology and management of neotropical migratory birds: a synthesis and review of critical issues. Oxford University Press, New York, NY.

PETIT, D. R., K. E. PETIT, AND T. C. GRUBB, JR. 1985. On atmospheric moisture as a factor influencing the distribution of breeding birds in temperate deciduous forest. Wilson Bulletin 97:88–96.

PETIT, D. R., K. E. PETIT, AND L. J. PETIT. 1990. Geographic variation in the foraging ecology of North American insectivorous birds. Studies in Avian Biology 13:254–263.

PETIT, L. J., D. R. PETIT, AND T. E. MARTIN. 1995.

Managing forests for migratory birds: looking past the trees to see the forest. Wildlife Society Bulletin 23:420–429.

PHILLIPS, A. R. 1975. The migrations of Allen's and other hummingbirds. Condor 77:196–205.

PIPER, W. H. 1997. Social dominance in birds: early findings and new horizons. Current Ornithology 14: 125–187.

PIPER, W. H., AND R. H. WILEY. 1990. The relationship between social dominance, subcutaneous fat, and annual survival in wintering White-throated Sparrows (*Zonotrichia albicollis*). Behavioral Ecology and Sociobiology 26:201–208.

PLACE, A. R., AND E. W. STILES. 1992. Living off the wax of the land: bayberries and Yellow-rumped Warblers. Auk 109:334–345.

POULIN, B., G. LEFEBVRE, AND R. MCNEIL. 1994. Diets of land birds from northeastern Venezuela. Condor 96:354–367.

POWER, D. M. 1971. Warbler ecology: diversity, similarity and seasonal differences in habitat segregation. Ecology 52:435–443.

PRESS, D., D. F. DOAK, AND P. STEINBERG. 1996. The role of local government in the conservation of rare species. Conservation Biology 10:1538–1548.

PULLIAM, H. R. 1973. On the advantages of flocking. Journal of Theoretical Biology 38:419–422.

QUAY, W. B. 1985. Sperm release in migrating wood-warblers (Parulinae) nesting at higher latitudes. Wilson Bulletin 97:283–295.

QUAY, W. B. 1989. Insemination of Tennessee Warblers during spring migration. Condor 91:660–670.

RAITT, R. J., AND S. L. PIMM. 1976. Dynamics of bird communities in the Chihuahua Desert, New Mexico. Condor 78:427–442.

RALPH, C. J. 1971. An age differential of migrants in coastal California. Condor 73:243–246.

RALPH, C. J. 1978. Disorientation and possible fate of young passerine coastal migrants. Bird-Banding 49: 237–247.

RALPH, C. J. 1981. Age ratios and their possible use in determining autumn routes of passerine migrants. Wilson Bulletin 93:164–188.

RAMOS, M. A. 1988. Eco-evolutionary aspects of bird movements in the northern neotropical region. Acta XIX Congressus Internationalis Ornithologici 19: 251–293.

RAPPOLE, J. H., AND M. V. MCDONALD. 1994. Cause and effect in population declines of migratory birds. Auk 111:652–660.

RAPPOLE, J. H., E. S. MORTON, T. E. LOVEJOY, III, AND J. L. RUOS. 1983. Nearctic avian migrants in the Neotropics. U.S. Fish and Wildlife Service, Washington, DC.

RAPPOLE, J. H., AND M. A. RAMOS. 1994. Factors affecting migratory bird routes over the Gulf of Mexico. Bird Conservation International 4:251–262.

RAPPOLE, J. H., AND D. W. WARNER. 1976. Relationships between behavior, physiology and weather in avian transients at a migration stopover site. Oecologia 26:193–212.

RAPPOLE, J. H., AND D. W. WARNER. 1980. Ecological aspects of migrant bird behavior in Veracruz, Mexico. Pp. 353–393 *in* A. Keast and E. S. Morton (editors). Migrant birds in the Neotropics: ecology, behavior, distribution, and conservation. Smithsonian Institution Press, Washington, D.C.

RAYNER, J. 1988. Form and function in avian flight. Current Ornithology 5:1–66.

RAYNER, J. 1990. The mechanics of flight and bird migration performance. Pp. 283–299 *in* E. Gwinner (editor). Bird migration: the physiology and ecophysiology. Springer-Verlag, Berlin, FRG.

REAL, L. 1980. Fitness, uncertainty, and the role of diversification in evolution and behavior. American Naturalist 115:623–638.

REAL, L., AND T. CARACO. 1986. Risk and foraging in stochastic environments. Annual Review of Ecology and Systematics 17:371–390.

REESE, W. A. 1984. Mosses of the Gulf South. Louisiana State University Press, Baton Rouge, LA.

REMSEN, J. V., JR., AND S. K. ROBINSON. 1990. A classification scheme for foraging behavior of birds in terrestrial habitats. Studies in Avian Biology 13: 144–160.

REY, P. J. 1995. Spatio-temporal variation in fruit and frugivorous bird abundance in olive orchards. Ecology 76:1625–1635.

RICE, J., B. W. ANDERSON, AND R. D. OHMART. 1980. Seasonal habitat selection by birds in the lower Colorado River Valley. Ecology 61:1402–1411.

RICE, W. R. 1989. Analyzing tables of statistical tests. Evolution 43:223–225.

RICHARDSON, W. J. 1978. Timing and amount of bird migration in relation to weather: a review. Oikos 30: 224–272.

RICHARDSON, W. J. 1990. Timing of bird migration in relation to weather: updated review. Pp. 78–101 *in* E. Gwinner (editor). Bird migration: the physiology and ecophysiology. Springer-Verlag, Berlin, FRG.

ROBBINS, C. S. 1979. Effects of forest fragmentation on bird populations. Pp. 198–212 *in* R. M. DeGraaf and K. E. Evans (compilers). Management of north central and northeastern forests for nongame birds. USDA Forest Service Gen. Tech. Rep. NC-51. USDA Forest Service North Central Experiment Station, St. Paul, MN.

ROBBINS, C. S., D. BYSTRAK, AND P. H. GEISSLER. 1986. The Breeding Bird Survey: its first fifteen years, 1965–1979. U.S. Fish and Wildlife Service Resource Publication 157, Washington, DC.

ROBBINS, C. S., D. K. DAWSON, AND B. A. DOWELL. 1989a. Habitat area requirements of breeding forest birds of the middle Atlantic States. Wildlife Monograph 103. The Wildlife Society, Washington, DC.

ROBBINS, C. S., J. R. SAUER, R. S. GREENBERG, AND S. DROEGE. 1989b. Population declines in North American birds that migrate to the Neotropics. Proceedings of the National Academy of Sciences (USA) 86:7658–7662.

ROBERTS, E. P., JR., AND P. D. WEIGL. 1984. Habitat preference in the Dark-eyed Junco (*Junco hyemalis*): the role of photoperiod and dominance. Animal Behavior 32:709–714.

ROBINSON, S. K., AND R. T. HOLMES. 1982. Foraging behavior of forest birds: the relationships among search tactics, diet, and habitat structure. Ecology 63:1918–1931.

ROBINSON, S. K., AND R. T. HOLMES. 1984. Effects of

plant species and foliage structure on the foraging behavior of forest birds. Auk 101:672–684.

ROBINSON, S. K., F. R. THOMPSON, III, T. M. DONOVAN, D. R. WHITEHEAD, AND J. FAABORG. 1995. Regional forest fragmentation and the nesting success of migratory birds. Science 267:1987–1990.

RODIEK, J. E., AND E. G. BOLEN (EDITORS). 1991. Wildlife and habitats in managed landscapes. Island Press, Washington, DC.

ROGERS, D. T., AND E. P. ODUM. 1966. A study of autumnal postmigrant weights and vernal fattening of North American migrants in the tropics. Wilson Bulletin 78:425–433.

ROOT, T. 1988. Atlas of wintering North American birds: an analysis of Christmas bird count data. University of Chicago Press, Chicago, IL.

ROSENBERG, K. V., R. D. OHMART, AND B. W. ANDERSON. 1982. Community organization of riparian breeding birds: response to an annual resource peak. Auk 99:260–274.

ROSENBERG, K. V., R. D. OHMART, W. C. HUNTER, AND B. W. ANDERSON. 1991. Birds of the lower Colorado River valley. University of Arizona Press, Tucson, AZ.

ROTH, R. R. 1976. Spatial heterogeneity and bird species diversity. Ecology 57:773–782.

RUDEBECK, G. 1950. The choice of prey and modes of hunting predatory birds with special reference to their selective effects. Oikos 2:65–88.

RUDEBECK, G. 1951. The choice of prey and modes of hunting predatory birds with special reference to their selective effects. Oikos 3:200–231.

RUDY, C. 1971. Age ratios in fall warblers. Inland Bird Banding News 43:97–101.

RUNDLE, W. D., AND L. H. FREDRICKSON. 1981. Managing seasonally flooded impoundments for migrant rails and shorebirds. Wildlife Society Bulletin 9:80–87.

RUSLING, W. J. 1936. Unpublished report to the National Association of Audubon Societies, New York, NY.

RUSSELL, R. J., AND H. V. HOWE. 1935. Cheniers of southwestern Louisiana. The Geographic Review 25:449–461.

RUSSELL, R. W., F. L. CARPENTER, M. A. HIXON, AND D. C. PATON. 1994. The impact of variation in stopover habitat quality on migrant Rufous Hummingbirds. Conservation Biology 8:483–490.

RYBCZYNSKI, R., AND D. K. RIKER. 1981. A temperate species-rich assemblage of migrant frugivorous birds. Auk 98:176–179.

SAAB, V. A., AND D. R. PETIT. 1992. Impact of pasture development on winter bird communities in Belize, Central America. Condor 94:66–71.

SAFRIEL, U. N., AND D. LAVEE. 1988. Weight changes of cross-desert migrants at an oasis—do energetic considerations alone determine the length of stopover? Oecologia 76:611–619.

SANDBERG, R., AND F. R. MOORE. 1996. Migratory orientation of Red-eyed Vireos, *Vireo olivaceus,* in relation to energetic condition and ecological context. Behavioral Ecology and Sociobiology 39:1–10.

SANDBERG, R., J. PETTERSSON, AND T. ALERSTAM. 1988. Shifted magnetic fields lead to deflected and axial orientation of migrating Robins, *Erithacus rubecula,* at sunset. Animal Behavior 36:877–887.

SANDBERG, R., J. PETTERSSON, AND K. PERSSON. 1991. Migratory orientation of free-flying Robins *Erithacus rubecula* and Pied Flycatchers *Ficedula hypoleuca*: release experiments. Ornis Scandinavica 22: 1–11.

SAS INSTITUTE. 1990. SAS/STAT user's guide, ver. 6, 4th ed., vol. 2. SAS Institute, Inc., Cary, NC.

SCHMITT, J. 1994. Reaction norms of morphological and life history traits to light availability in *Impatiens capensis.* Evolution 47:1654–1668.

SCHNEIDER, K. J. 1984. Dominance, predation, and optimal foraging in White-throated Sparrow flocks. Ecology 65:1820–1827.

SCHOENER, T. W. 1974. Competition and the form of habitat shift. Theoretical Population Biology 6:265–307.

SCHWEITZER, S. J., D. M. FINCH, AND D. M. LESLIE, JR. 1996. Reducing impacts of brood parasitism by Brown-headed Cowbirds on riparian-nesting migratory songbirds. Pp. 267–276 *in* D. W. Shaw and D. M. Finch (technical coordinators). Desired future conditions for southwestern riparian ecosystems: bringing interests and concerns together. USDA Forest Service Gen. Tech. Rep. RM-272. USDA Forest Service Rocky Mountain Forest and Range Experiment Station, Fort Collins, CO.

SCURLOCK, D. 1995. Environmental history. Pp. 12–28 *in* D. M. Finch and J. A. Tainter (editors). Ecology and sustainability of the Middle Rio Grande Basin. USDA Forest Service Gen. Tech. Rep. RM-268. USDA Forest Service Rocky Mountain Forest and Range Experiment Station, Fort Collins, CO.

SCURLOCK, D. 1998. From the Rio to the Sierra: an environmental history of the Middle Rio Grande Basin. USDA Forest Service Gen. Tech. Rep. RMRS-5. USDA Forest Service Rocky Mountain Forest and Range Experiment Station, Fort Collins, CO.

SEALY, S. G. 1988. Aggressiveness in migrating Cape May Warblers: defense of an aquatic food source. Condor 90:271–274.

SEALY, S. G. 1989. Defense of nectar resources by migrating Cape May warblers. Journal of Field Ornithology 60:89–93.

SEEL, D. C. 1977. Migration of the northwestern European population of the Cuckoo *Cuculus canorus,* as shown by ringing. Ibis 119:309–322.

SELANDER, R. K. 1966. Sexual dimorphism and differential niche utilization in birds. Condor 68:113–151.

SHAFER, C. L. 1990. Nature reserves: island theory and conservation practice. Smithsonian Institution Press, Washington, DC.

SHERRY, T. W. 1984. Comparative dietary ecology of sympatric insectivorous neotropical flycatchers (Tyrannidae). Ecological Monographs 54:313–338.

SHERRY, T. W. 1990. When are birds dietarily specialized? Distinguishing ecological from evolutionary approaches. Studies in Avian Biology No. 13:337–352.

SHERRY, T. W., AND R. T. HOLMES. 1985. Dispersion patterns and habitat responses of birds in northern hardwood forests. Pp. 283–310 *in* M. Cody (editor). Habitat selection in birds. Academic Press, San Diego, CA.

SHERRY, T. W., AND R. T. HOLMES. 1989. Age-specific social dominance affects habitat use by breeding American Redstarts (*Setophaga ruticilla*): a removal experiment. Behavioral Ecology and Sociobiology 25:327–333.

SHERRY, T. W., AND R. T. HOLMES. 1992. Population fluctuations in a long-distance neotropical migrant: demographic evidence for the importance of breeding season events in the American Redstart. Pp. 431–442 *in* J. M. Hagan, III, and D. W. Johnston (editors). Ecology and conservation of neotropical migrant landbirds. Smithsonian Institution Press, Washington, D.C.

SHERRY, T. W., AND R. T. HOLMES. 1993. Are populations of neotropical migrant birds limited in summer or winter? Implications for management. Pp. 47–57 *in* D. M. Finch and P. W. Stangel (editors). Status and management of neotropical migratory birds. USDA Forest Service Gen. Tech. Rep. RM-229. USDA Forest Service Rocky Mountain Forest and Range Experiment Station, Ft. Collins, CO.

SHERRY, T. W., AND R. T. HOLMES. 1995. Summer versus winter limitation of populations: what are the issues and what is the evidence? Pp. 85–120 *in* T. E. Martin and D. M. Finch (editors). Ecology and management of neotropical migratory birds: a synthesis and review of critical issues. Oxford University Press, New York, NY.

SICCAMA, T. G. 1971. Presettlement and present forest vegetation in northern Vermont with special reference to Chittenden County. American Midland Naturalist 85:153–161.

SIMBERLOFF, D., AND J. COX. 1987. Consequences and costs of conservation corridors. Conservation Biology 1:63–71.

SIMONS, D., AND F. BAIRLEIN. 1990. The significance of seasonal frugivory in migratory Garden Warblers, *Sylvia borin*. Journal für Ornithologie 131:381–402.

SKAGEN, S. K., C. P. MELCHER, W. H. HOWE, AND F. L. KNOPF. 1998. Comparative use of riparian corridors and oases by migrating birds in southeast Arizona. Conservation Biology 12:896–909.

SMITH, R., M. HAMAS, M. DALLMAN, AND D. EWERT. 1998. Spatial variation in foraging of the Black-throated Green Warbler along the shoreline of northern Lake Huron. Condor 100:474–484.

SNOW, D. W. 1971. Evolutionary aspects of fruit-eating by birds. Ibis 113:194–202.

SOULÉ, M. E. 1991. Land use planning and wildlife maintenance: guidelines for conserving wildlife in an urban landscape. Journal of the American Planning Association 57:313–323.

SPEARING, D. 1995. Roadside geology of Louisiana. Mountain Press Publishing Co., Missoula, MT.

SPINA, F., AND E. M. BEZZI. 1990. Autumn migration and orientation of the Sedge Warbler (*Acrocephalus schoenobaenus*) in northern Italy. Journal für Ornithologie 131:429–438.

SPRUNT, A., IV. 1975. Habitat management implications of migration. Pp. 81–86 *in* D. R. Smith (technical coordinator). Proceedings of the symposium on management of forest and range habitats for nongame birds. USDA Forest Service Gen. Tech. Rep. WO-1. USDA Forest Service, Washington, D.C.

STEPHENS, D. W., AND E. L. CHARNOV. 1982. Optimal foraging: some simple stochastic models. Behavioral Ecology and Sociobiology 10:251–263.

STEVENS, L. E., B. T. BROWN, J. M. SIMPSON, AND R. R. JOHNSON. 1977. The importance of riparian habitat to migrating birds. Pp. 156–164 *in* R. R. Johnson and D. A. Jones, Jr. (technical coordinators). Importance, preservation and management of riparian habitat: a symposium. USDA Forest Service Gen. Tech. Rep. RM-43. USDA Forest Service Rocky Mountain Forest and Range Experiment Station, Fort Collins, CO.

STEWART, R. M., L. R. MEWALDT, AND S. KAISER. 1974. Age ratios of coastal and inland fall migrant passerines in central California. Bird-Banding 45: 46–57.

STILES, E. W. 1980. Patterns of fruit presentation and seed dispersal in bird-disseminated woody plants in the eastern deciduous forest. American Naturalist 116:670–688.

STILES, F. G. 1980. Evolutionary implications of habitat relations between permanent and winter land birds in Costa Rica. Pp. 421–435 *in* A. Keast and E. S. Morton (editors). Migrant birds in the Neotropics: ecology, behavior, distribution, and conservation. Smithsonian Institution Press, Washington, D.C.

STILES, F. G., AND A. F. SKUTCH. 1989. A guide to the birds of Costa Rica. Cornell University Press, Ithaca, NY.

STOATE, C., AND S. J. MOREBY. 1995. Premigratory diet of trans-Saharan migrant passerines in the western Sahel. Bird Study 42:101–106.

STUTCHBURY, B. J. 1994. Competition for winter territories in a neotropical migrant: the role of age, sex and color. Auk 111:63–69.

SULLIVAN, K. A. 1988. Age-specific profitability and prey choice. Animal Behavior 36:613–615.

SVENSSON, S. E. 1978. Efficiency of two methods for monitoring bird population levels: breeding bird censuses contra counts of migrating birds. Oikos 30: 373–386.

SYKES, P. W., JR. 1986. Autumn land-bird migration on the barrier islands of northeastern North Carolina. Occasional Papers of the North Carolina Biological Survey 1986-11, Raleigh, NC.

SZARO, R. C., AND R. P. BALDA. 1979. Bird community dynamics in a ponderosa pine forest. Studies in Avian Biology 3:1–66.

TAYLOR, D. M. 1986. Effects of cattle grazing on passerine birds nesting in riparian habitat. Journal of Range Management 39:254–258.

TAYLOR, D. M., AND C. D. LITTLEFIELD. 1986. Willow Flycatcher and Yellow Warbler response to cattle grazing. American Birds 40:1169–1173.

TAYLOR, M. J., M. R. BYRNES, AND R. A. MCBRIDE. 1995. Sediment texture and composition changes along the southwest Louisiana coast: implications for sediment supply. Gulf Coast Association of Geological Societies Transactions 45:557–564.

TERBORGH, J. W. 1977. Bird species diversity on an Andean elevation gradient. Ecology 58:1007–1019.

TERBORGH, J. W. 1980. The conservation status of neotropical migrants: present and future. Pp. 21–30 *in* A. Keast and E. S. Morton (editors). Migrant birds in the Neotropics: ecology, behavior, distribution, and conservation. Smithsonian Institution Press, Washington, D.C.

TERRILL, S. B. 1987. Social dominance and migratory restlessness in the Dark-eyed Junco (*Junco hyemalis*). Behavioral Ecology and Sociobiology 21:1–11.

TERRILL, S. B. 1988. The relative importance of ecological factors in bird migration. Acta XIX Congressus Internationalis Ornithologici 19:2180–2190.

TERRILL, S. B. 1990. Food availability, migratory behavior and population dynamics of terrestrial birds during the non-reproductive season. Studies in Avian Biology 13:438–443.

TERRILL, S. B., AND R. D. OHMART. 1984. Facultative extension of fall migration by Yellow-rumped Warblers (*Dendroica coronata*). Auk 101:427–438.

TEXAS BUREAU OF ECONOMIC GEOLOGY. 1976. Environmental geologic atlas of the Texas coastal zone-Galveston bay area. University of Texas Press, Austin, TX.

THE BARRED OWL. 1996a. Vol. XXII; No. 9. Newsletter of the Baton Rouge Audubon Society, Baton Rouge, LA.

THE BARRED OWL. 1996b. Vol. XXII; No. 12. Newsletter of the Baton Rouge Audubon Society, Baton Rouge, LA.

THOMAS, D. K. 1979. Figs as a fruit source of migrating Garden Warblers in southern Portugal. Bird Study 26:187–191.

THOMPSON, B. C., D. A. LEAL, AND R. A. MEYER. 1994. Bird community composition and habitat importance in the Rio Grande system of New Mexico with emphasis on neotropical migrant birds. Research Completion Report. Cooperative Agreement 14-16-0009-1572, U.S. Fish and Wildlife Service and National Biological Survey, Albuquerque, NM.

THOMPSON, J. D. 1991. Phenotypic plasticity as a component of evolutionary change. Trends in Ecology and Evolution 6:246–249.

THOMPSON, W. L. 1960. Agnostic behavior in the House Finch: 2: Factors in aggressiveness and sociality. Condor 62:378–402.

TRAMER, E. J., AND T. R. KEMP. 1979. Diet-correlated variations in social behavior of wintering Tennessee Warblers. Auk 96:186–187.

TURNER, M. G., AND R. H. GARDNER (EDITORS). 1991. Quantitative methods in landscape ecology: the analysis and interpretation of landscape heterogeneity. Springer Verlag, New York, NY.

U.S. DEPARTMENT OF THE INTERIOR. 1978. The Atchafalaya, America's greatest river swamp: proposal to establish the Atchafalaya fish, wildlife, and multi-use area. US Fish and Wildlife Service Proposal, Lafayette, LA.

U.S. DEPARTMENT OF THE INTERIOR. 1993. Meeting our coastal challenges. Special Report, US Fish and Wildlife Service, Branch of Coastal and Wetland Resources, Arlington, VA.

U.S. GENERAL SERVICES ADMINISTRATION. 1993. Summary report of real property owned by the United States throughout the world as of September 30, 1990. Congressional Information Services, Bethesda, MD.

VALONE, T. J. 1992. Information for patch assessment: a field investigation with Black-chinned Hummingbirds. Behavioral Ecology 3:211–222.

VAN DERSAL, W. R. 1939. Birds that feed on Russian-olive. Auk 56:483–484.

VEIGA, J. P. 1986. Settlement and fat accumulation by migrant Pied Flycatchers in Spain. Ringing and Migration 7:85–98.

VICKERY, P. D., M. L. HUNTER, JR., AND S. M. MELVIN. 1994. Effects of habitat area on the distribution of grassland birds in Maine. Conservation Biology 8: 1087–1097.

VINES, R. A. 1960. Trees, shrubs, and woody vines of the Southwest. University of Texas Press, Austin, TX.

WAITE, T. A. 1987. Vigilance in the White-breasted Nuthatch: effects of social dominance and sociality. Auk 104:429–434.

WALTER, H. 1979. Eleonora's Falcon: adaptations to prey and habitat in a social raptor. University of Chicago Press, Chicago, IL.

WARD, S. A. 1987. Optimal habitat selection in time-limited dispersers. American Naturalist 129:568–579.

WATTS, B. D., AND S. E. MABEY. 1993. Spatio-temporal patterns of landbird migration on the lower Delmarva Peninsula. Annual report to the Virginia Department of Environmental Quality, Richmond, VA.

WATTS, B. D., AND S. E. MABEY. 1994. Migratory landbirds on the lower Delmarva: habitat selection and geographic distribution. Report No. NA37OZ0360-01, Virginia Department of Environmental Quality Coastal Resource Management Program, Richmond, VA.

WAUER, R. H. 1977. Significance of Rio Grande riparian systems upon the avifauna. Pp. 165–174 *in* R. R. Johnson and D. A. Jones, Jr. (technical coordinators). Importance, preservation and management of riparian habitat: a symposium. USDA Forest Service Gen. Tech. Rep. RM-43. USDA Forest Service Rocky Mountain Forest and Range Experiment Station, Fort Collins, CO.

WEATHERS, W. W., AND K. A. SULLIVAN. 1991. Foraging efficiency of parent juncos and their young. Condor 93:346–353.

WECKER, S. C. 1964. Habitat selection. American Scientist 211:109–116.

WEGNER, J. F., AND G. MERRIAM. 1979. Movements by birds and small mammals between a wood and adjoining farmland habitats. Journal of Applied Ecology 16:349–357.

WEISBROD, A. R., C. J. BURNETTE, J. G. TURNER, AND D. W. WARNER. 1993. Migrating birds at a stopover site in the Saint Croix River valley. Wilson Bulletin 105:265–284.

WESTMAN, W. E. 1990. Park management of exotic plant species: problems and issues. Conservation Biology 4:251–260.

WESTNEAT, D. F., P. W. SHERMAN, AND M. L. MORTON. 1990. The ecology and evolution of extra-pair copulations in birds. Current Ornithology 7:331–369.

WHEELWRIGHT, N. T. 1988. Seasonal changes in food preferences of American Robins in captivity. Auk 105:374–378.

WHEELWRIGHT, N. T., W. A. HABER, K. G. MURRAY, AND C. GUINDON. 1984. Tropical fruit-eating birds and their food plants: a survey of a Costa Rican lower montane forest. Biotropica 16:173–192.

WHELAN, C. 1989. Avian foliage structure preferences for foraging and the effect of prey biomass. Animal Behavior 38:839–846.

WHITCOMB, R. F. 1977. Island biogeography and habitat islands of eastern forest. American Birds 31:3–5.

WHITCOMB, R. F., J. F. LYNCH, P. A. OPLER, AND C. S. ROBBINS. 1976. Island biogeography and conservation: strategy and limitations. Science 193:1030–1032.

WHITCOMB, R. F., C. S. ROBBINS, J. F. LYNCH, B. L. WHITCOMB, M. K. KLIMKIEWICZ, AND D. BYSTRAK. 1981. Effects of forest fragmentation on avifauna of the eastern deciduous forest. Pp. 125–205 *in* R. L. Burgess and D. M. Sharpe (editors). Forest island dynamics in man-dominated landscapes. Springer-Verlag, New York, NY.

WHITE, D. W. 1989. North American bird-dispersed fruits: ecological and adaptive significance of nutritional and structural traits. Ph.D. dissertation. Rutgers University, Piscatawy, NJ.

WHITE, D. W., AND E. W. STILES. 1990. Co-occurrence of foods in the stomachs and feces of fruit-eating birds. Condor 92:291–303.

WIEDNER, D., AND P. KERLINGER. 1990. Economics of birding: a national survey of active birders. American Birds 44:209–213.

WIEDNER, D. S., P. KERLINGER, D. A. SIBLEY, P. HOLT, J. HOUGH, AND R. CROSSLEY. 1992. Visible morning flights of neotropical landbird migrants at Cape May, New Jersey. Auk 109:500–510.

WIENS, J. A. 1981. Scale problems in avian censusing. Studies in Avian Biology 6:513–521.

WIENS, J. A. 1989. The ecology of bird communities, vol. 2. Cambridge University Press, Cambridge, UK.

WIGLEY, T. B., AND J. M. SWEENEY. 1993. Cooperative partnerships and the role of private landowners. Pp. 39–44 *in* D. M. Finch and P. W. Stangel (editors). Status and management of neotropical migratory birds. USDA Forest Service Gen. Tech. Rep. RM-229. USDA Forest Service Rocky Mountain Forest and Range Experiment Station, Ft. Collins, CO.

WILCOVE, D. S. 1988. Changes in the avifauna of the Great Smoky Mountains: 1947–1983. Wilson Bulletin 100:256–271.

WILCOVE, D. S., AND S. K. ROBINSON. 1990. The impact of forest fragmentation on bird communities in eastern North America. Pp. 319–331 *in* A. Keast (editor). Biogeography and ecology of forest bird communities. Academic Publishing, The Hague, Netherlands.

WILCOVE, D. S., AND J. W. TERBORGH. 1984. Patterns of population decline in birds. American Birds 38: 10–13.

WILCOVE, D. S., AND R. F. WHITCOMB. 1983. Gone with the trees. Natural History 92:82–91.

WILLIAMSON, P., AND L. GRAY. 1975. Foraging behavior of the Starling (*Sturnus vulgaris*) in Maryland. Condor 77:84–89.

WILLIS, E. O. 1980. Ecological roles of migratory and resident birds on Barro Colorado Island, Panama. Pp. 205–225 *in* A. Keast and E. S. Morton (editors). Migrant birds in the Neotropics: ecology, behavior, distribution, and conservation. Smithsonian Institution Press, Washington, D.C.

WILLSON, M. F. 1974. Avian community organization and habitat structure. Ecology 55:1017–1029.

WILLSON, M. F. 1991. Birds and fruits: how does this mutualism matter? Acta XX Congressus Internationalis Ornithologici 20:1630–1635.

WILLSON, M. F., AND S. W. CAROTHERS. 1979. Avifauna of habitat islands in the Grand Canyon. Southwestern Naturalist 24:563–576.

WILLSON, M. F., E. A. PORTER, AND R. S. CONDIT. 1982. Avian frugivore activity in relation to forest light gaps. Caribbean Journal of Science 18:1–4.

WILLSON, M. F., AND J. N. THOMPSON. 1982. Phenology and ecology of color in bird-dispersed fruits, or why some fruits are red when they are "green". Canadian Journal of Botany 60:701–713.

WILLSON, M. F., AND C. J. WHELAN. 1990. The evolution of fruit color in fleshy-fruited plants. American Naturalist 136:790–809.

WILSON, E. O. 1984. Biophilia. Harvard University Press, Cambridge, MA.

WILZ, K. J., AND V. GIAMPA. 1978. Habitat use by Yellow-rumped Warblers at the northern extremities of their winter range. Wilson Bulletin 90:566–574.

WINKER, K. D. 1995. Autumn stopover on the Isthmus of Tehuantepec by woodland nearctic-neotropical migrants. Auk 112:690–700.

WINKER, K. D. 1995. Habitat selection in woodland nearctic-neotropical migrants on the Isthmus of Tehuantepec I. Autumn migration. Wilson Bulletin 107:26–39.

WINKER, K. D., J. H. RAPPOLE, AND M. A. RAMOS. 1995. The use of movement data as an assay of habitat quality. Oecologia 101:211–216.

WINKER, K. D., D. W. WARNER, AND A. R. WEISBROD. 1992a. The Northern Waterthrush and Swainson's Thrush as transients at a temperate inland stopover site. Pp. 384–402 *in* J. M. Hagan, III, and D. W. Johnston (editors). Ecology and conservation of neotropical migrant landbirds. Smithsonian Institution Press, Washington, D.C.

WINKER, K. D., D. W. WARNER, AND A. R. WEISBROD. 1992b. Daily mass gains among woodland migrants at an inland stopover site. Auk 109:853–862.

WINKER, K. D., D. W. WARNER, AND A. R. WEISBROD. 1992c. Migration of woodland birds at a fragmented inland stopover site. Wilson Bulletin 104:580–598.

WOLDA, H. 1978. Seasonal fluctuations in rainfall, food and abundance of tropical insects. Journal of Animal Ecology 47:369–381.

WOLDA, H. 1990. Food availability for an insectivore and how to measure it. Studies in Avian Biology 13: 29–37.

WOLDA, H. 1988. Insect seasonality: Why? Annual Review of Ecology and Systematics 19:1–18.

WOOD, B. 1992. Yellow Wagtail *Motacilla flava* migration from west Africa to Europe: pointers towards a conservation strategy for migrants on passage. Ibis 134 (suppl. 1):66–76.

WOODREY, M. S. 1995. Stopover behavior and age-specific ecology of neotropical passerine migrant landbirds during autumn along the northern coast of the Gulf of Mexico. Ph.D. dissertation. University of Southern Mississippi, Hattiesburg, MS.

WOODREY, M. S., AND C. R. CHANDLER. 1997. Age-related timing of migration: geographic and interspecific patterns. Wilson Bulletin 109:52–67.

WOODREY, M. S., AND F. R. MOORE. 1997. Age-related differences in the stopover of fall landbird migrants on the coast of Alabama. Auk 114:695–707.

WOZNIAK, F. E. 1995. Human ecology and ethnology. Pp. 29–51 *in* D. M. Finch and J. A. Tainter (editors). Ecology and sustainability of the Middle Rio Grande Basin. USDA Forest Service Gen. Tech. Rep. RM-268. USDA Forest Service Rocky Mountain Forest and Range Experiment Station, Fort Collins, CO.

WOZNIAK, F. E. 1998. Irrigation in the Rio Grande Valley, New Mexico: a study and annotated bibliography of the development of irrigation systems. USDA Forest Service RMRS-P-2. USDA Forest Service Rocky Mountain Research Station, Fort Collins, CO.

WUNDERLE, J. M., JR. 1991. Age-specific foraging proficiency in birds. Current Ornithology 8:273–324.

YAHNER, R. H. 1983. Seasonal dynamics, habitat relationships, and management of avifauna in farmstead shelterbelts. Journal of Wildlife Management 47:85–104.

YARBROUGH, C. G., AND D. W. JOHNSTON. 1965. Lipid deposition in wintering and premigratory Myrtle Warblers. Wilson Bulletin 77:175–191.

YDENBERG, R. C., AND L. M. DILL. 1986. The economics of fleeing from predators. Advances in the Study Behavior 16:229–249.

YONG, W., AND D. M. FINCH. 1996. Landbird species composition and relative abundance during migration along the Rio Grande. Pp. 77–92 *in* D. W. Shaw and D. M. Finch (technical coordinators). Desired future conditions for southwestern riparian ecosystems: bringing interests and concerns together. USDA Forest Service Gen. Tech. Rep. RM-272. USDA Forest Service Rocky Mountain Forest and Range Experiment Station, Fort Collins, CO.

YONG, W., AND D. M. FINCH. 1997a. Migration of the Willow Flycatcher along the middle Rio Grande. Wilson Bulletin 109:253–268.

YONG, W., AND D. M. FINCH. 1997b. Population trends of migratory landbirds along the middle Rio Grande. Southwestern Naturalist 42:132–147.

YONG, W., D. M. FINCH, AND S. W. COX. 1995. Landbird migration along the middle Rio Grande: summary of banding data from spring and fall 1994. New Mexico Ornithological Society Bulletin 23(3): 64–77.

YONG, W., D. M. FINCH, F. R. MOORE, AND J. F. KELLY. 1998. Stopover ecology and habitat use of migratory Wilson's Warblers. Auk 115:829–842.

YONG, W., AND F. R. MOORE. 1993. Relation between migratory activity and energetic condition among thrushes (Turdinae) following passage across the Gulf of Mexico. Condor 95:934–943.

YONG, W., AND F. R. MOORE. 1994. Flight morphology, energetic condition, and the stopover biology of migrating thrushes. Auk 111:683–692.

YONG, W., AND F. R. MOORE. 1997. Spring stopover of intercontinental migratory thrushes along the northern coast of the Gulf of Mexico. Auk 114:263–278.

ZAR, J. 1996. Biostatistical Analysis. 3rd edition. Prentice-Hall, Upper Saddle River, NJ.

ZEINER, D. C., W. F. LAUDENSLAYER, JR., K. E. MAYER, AND M. WHITE. 1990. California's wildlife. Vol. II: Birds. California Department of Game and Fish, Sacramento, CA.

ZIEGLER, H. P. 1976. Feeding behavior of the pigeon. Advances in the Study of Behavior 7:286–389.